Table of Contents

Scope

- This book is a good complement to prepare the ABYC certification in Gasoline engines. Additionally gives to reader the basic learning to diagnosis and repair of marine diesel engines
- This course **is not equivalent** to the certification granted by the engine manufacturers (Mercury, Volvo, Suzuki ,Yamaha, etc)
- We will follow the recommendations of **ABYC** Standards , the **USCG** regulations and the **CFR's** (code of federal regulations)

ABYC Certification

This book is a good complement for students interested to take the ABYC Gasoline Certification

Recognition

This book is the result of many years of practice and research about gasoline engines. Most of the items presented here are taken as reference, from research and articles of other authors. All I want to express is my gratitude because thanks to these works , this book will be useful for boat owners and marine engineers interested in gasoline engines service. In the same way some graphics are taken from public internet sites. To the people who have uploaded these images to the cloud , thank you very much for its contribution

Chapter 1
Inboard and Outboard Engines
General Features

Fuel Vs Efficiency

Planing hulls need much more power than displacement hulls and use more fuel per mile. However their ability to carry fuel is limited because their planing ability is affected by weight.

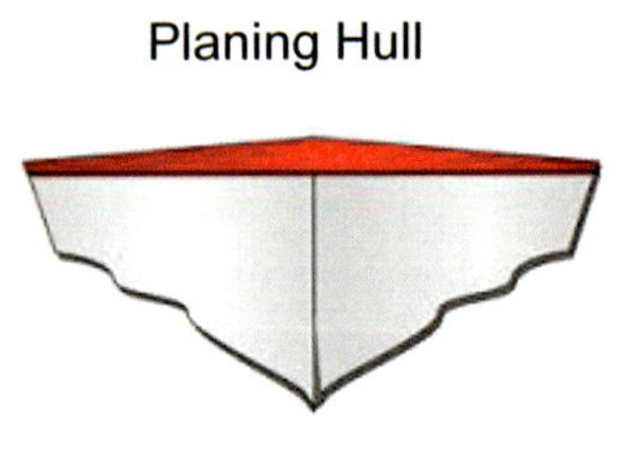

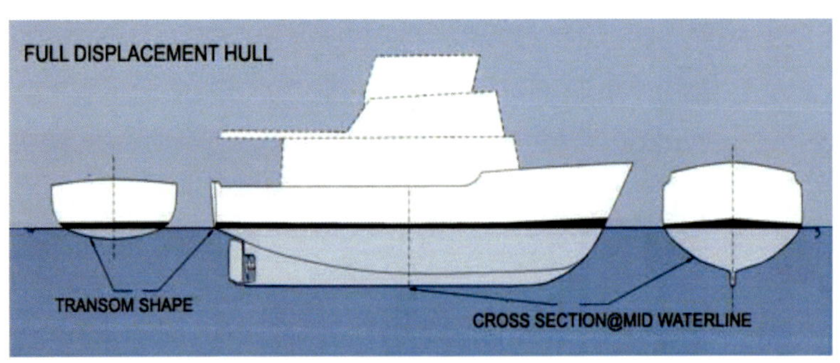

Planing Hulls

Planing hulls are called deep vee or modified deep vee with the vee shape along the keel less pronounced. Once a planing boat reaches the speed of approximately 12 knots/14 mph the hull actually lifts up – gets up on plane – to reduce its underwater hull surface and provide more speed

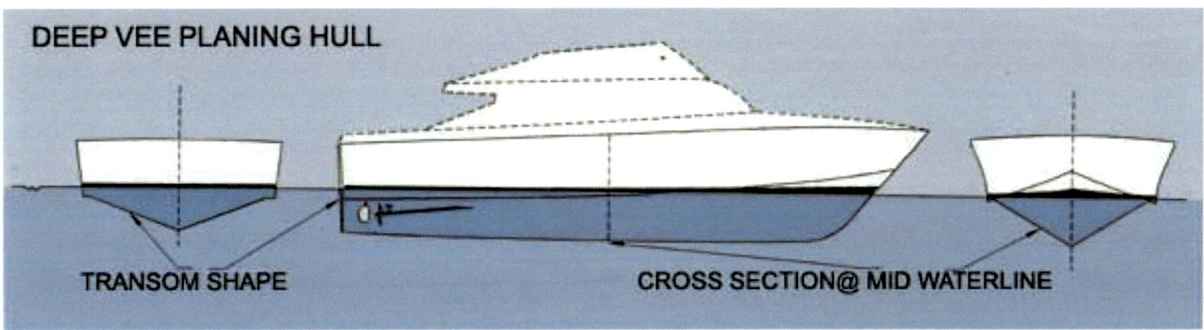

The Boat Thrust

As I said in my books and talking about in-board engine boats , what really produces the **thrust** depends of various factors such as :

- The type of hull
- The propeller selection
- The shaft and propeller configuration
- The engine output torque
- Other factors such as the transmission gear ratio , bottom hull surface, etc.

How much does the fuel weigh?

Gasoline weighs about 6.1 pounds per gallon. Diesel weights about 7.1 pounds per gallon

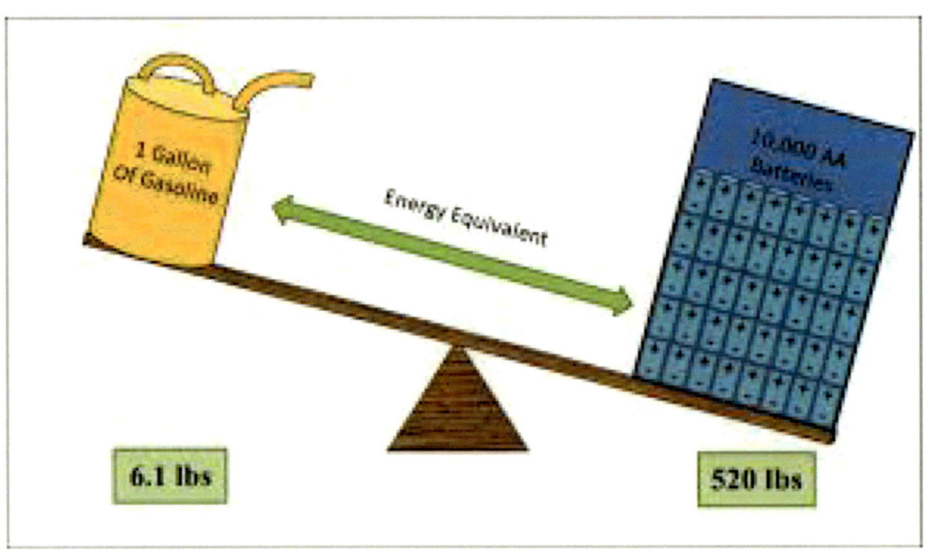

Gasoline Vs Diesel

When it comes to traversing great distances at highway speeds, the diesel engine's higher compression ratios and lean-burn combustion provide an efficiency that no gas engine can currently match

PETROL V/S DIESEL ENGINE

PETROL ENGINE (S.I.)	DIESEL ENGINE (C.I.)
▢ Otto cycle	▢ Diesel cycle
▢ Air – fuel mixture suction stroke	▢ Only air sucked during suction stroke
▢ spark plug is needed	▢ No spark plug needed
▢ C.R.=6-12	▢ C.R.= 14-22
▢ Low efficiency	▢ High efficiency
▢ Light weight	▢ Heavy
▢ Cheap	▢ Costly
▢ Less vibration & noise	▢ More
▢ Motor cycles, cars, light duty vehicles	▢ Trucks,buses,gensets

Thermal Efficiency

Over the diesels operating range, the average thermodynamic efficiency is highly superior . The diesel engine produce around of 15 or 30 percent better efficiency than a gas engine in the same conditions

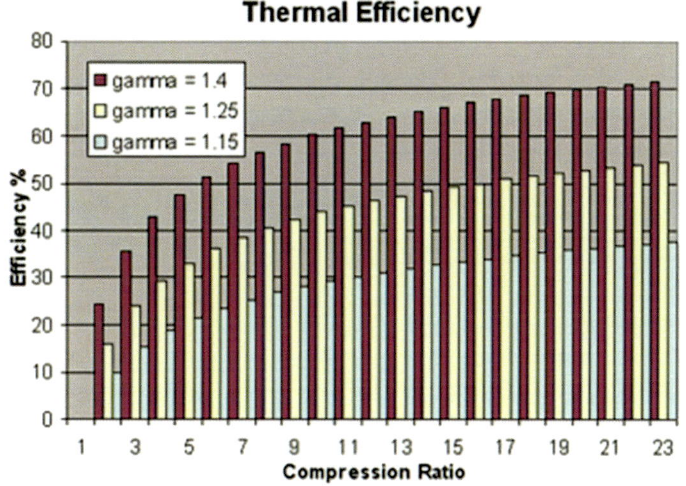

Fuel Vs Efficiency

Both gasoline and diesel engines use about 0.6 pound of fuel per horsepower per hour. On average, diesel fuel contains about 140,000 BTUs per gallon or 10 percent more energy than the same volume of gasoline

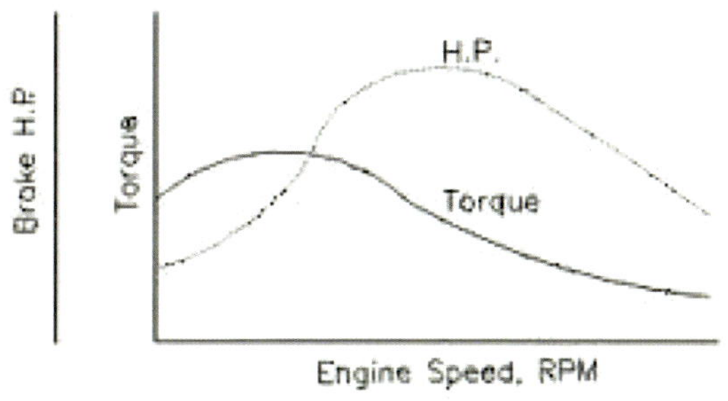

Marine Engines efficiency

An engine at cruising speed usually uses only about **two-thirds** of its maximum available horsepower. Most marine engines are designed to run continuously at between 60 and 75 percent of maximum speed. Diesels tend to be more toward the top of the range

Where does it go when it is consumed?

- Your engine uses the fuel you purchase in several ways.
  - 35 percent is given up to the atmosphere in heat
  - 25 percent is given up in heat and vibration absorbed by surrounding water
  - 10 percent is given up to overcome wave resistance
  - 6 percent to overcome wave formation and prop wash against the hull
  - 7 percent to overcome skin friction
  - 2 percent is wasted in friction at the propeller shaft
  - 1 percent to overcome air resistance
- This leaves **about 13-14 %** of the original energy to turn the propeller

How much do gasoline and diesel engines consume?

- **Diesel engines** consume about 1 gallon per hour for every 18 hp used. You can estimate the number of gallons consumed per hour by multiplying horsepower used by 0.055

- **Gasoline four stroke inboard engines** need about 1 gallon per hour for every 10 hp used. The number of gallons consumed per hour can be estimated by multiplying horsepower used by 0.100

Hull Speed Vs Fuel Consumption

Any attempt to force a displacement hull beyond its maximum theoretical hull speed brings exponentially higher fuel consumption for minimal gain in speed

Engine rpm	Speed-Knots	Fuel Burn-GPH
700	3.0	0.2
1000	4.2	0.4
1500	6.0	1.0
2000	8.5	2.2
2500	10.2	3.9
3000	14.7	5.6
3200	16.4	6.7
3400	17.7	7.7
3600	19.5	9.0
3800	20.4	11.0
4000	21.5	12.0

What Is An Engine?

Simply speaking, an engine is a group of related parts that are assembled in a way to convert energy into Torque that, in turn, can be harnessed to do work

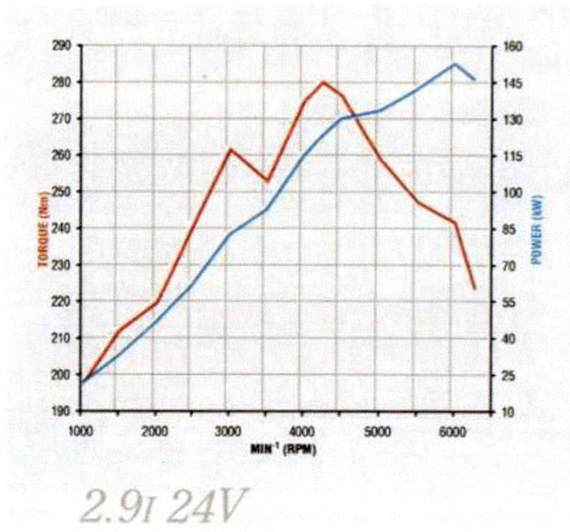

Marine Engines Classification

- **Inboard Engines**
  - Electric Motors
    - Alternating Current.
    - Direct Current
  - Diesel (2 &4 Strokes).
  - Gasoline
    - 4 Strokes

- **Outboard Engines**
  - 2 Strokes
  - 4 Strokes
  - Electric Motors.
    - DC 12/24 Volts
    - AC Motors

Inboard Gasoline Engines

- A **petrol engine** (known as a **gasoline engine** in North America) is an internal combustion engine with spark ignition

- It differs from a diesel engine in the method of mixing the fuel and air, and in the fact that it uses spark plugs to initiate the combustion process

Inboard Engines

- An inboard is a four-stroke automotive engine adapted for marine use. Inboard engines are mounted inside the hull's midsection or in front of the transom

- The engine turns a drive shaft that runs through the bottom of the hull and is attached to a propeller at the other end.

- Steering of most inboard vessels, except PWCs and jet-drive boats, is controlled by a rudder behind the propeller

In-Board And Out-board

- The marine in-board engines and out-board engines are classified according with the number of strokes per cycle

- A stroke is the movement of the piston in one direction, moving the piston from the top to the bottom of the cylinder is one stroke. A running internal combustion engine continually repeats a power cycle called: intake, compression, power and exhaust.

Four Stroke Gas Engine

The four cycles are done in two crankshaft revolutions. The head valves open and close, by the camshaft

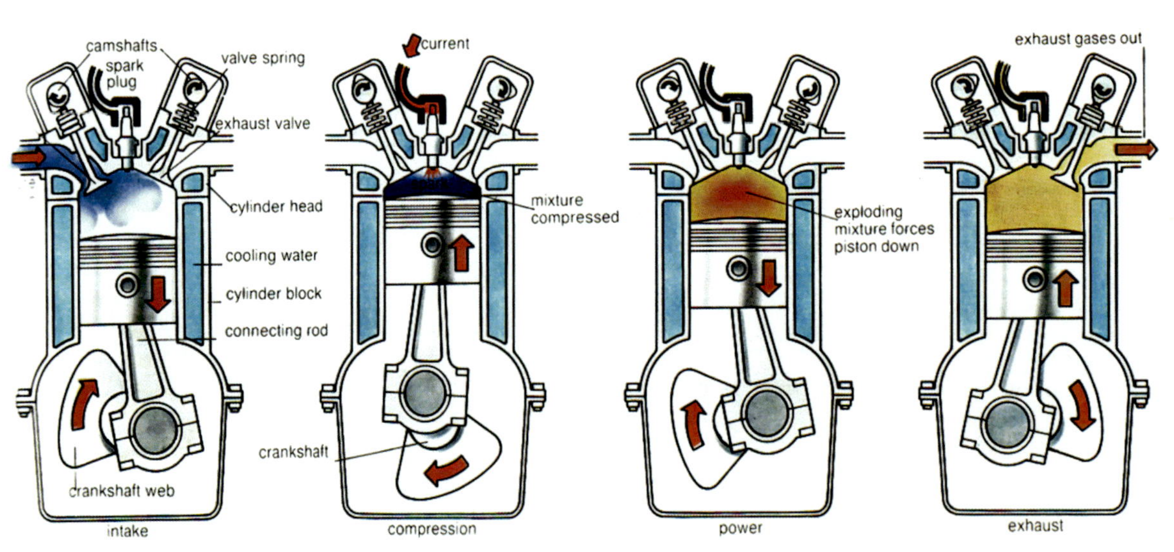

Outboard Engines
Two Stroke

Two Stroke Engines

A **two-stroke engine** is an internal combustion engine that completes the process cycle in one revolution of the crankshaft (an up stroke and a down stroke of the piston, compared to twice that number for a four-stroke engine)

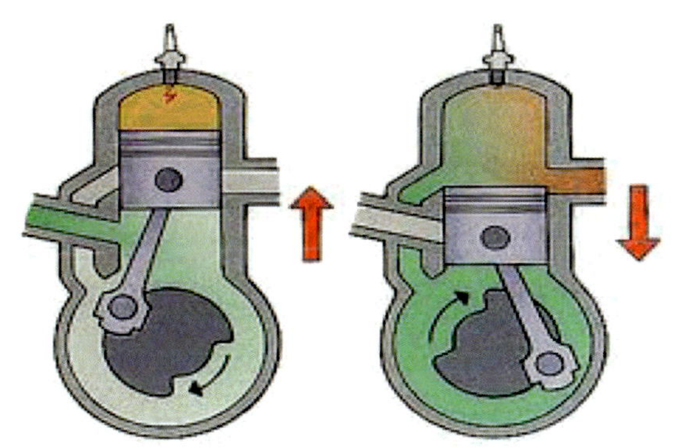

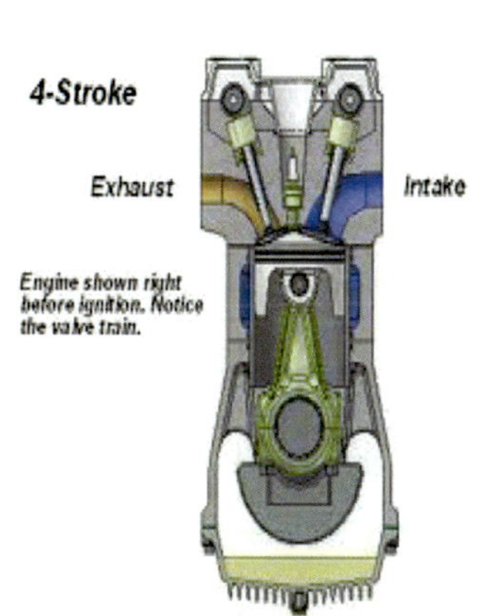

Two Stroke Cycle

"Stroke" refers to the movement of the piston in the engine. 2 Stroke means one stroke in each direction. A 2 stoke engine will have a compression stroke followed by an explosion of the compressed fuel. On the return stroke new fuel mixture is inserted into the cylinder

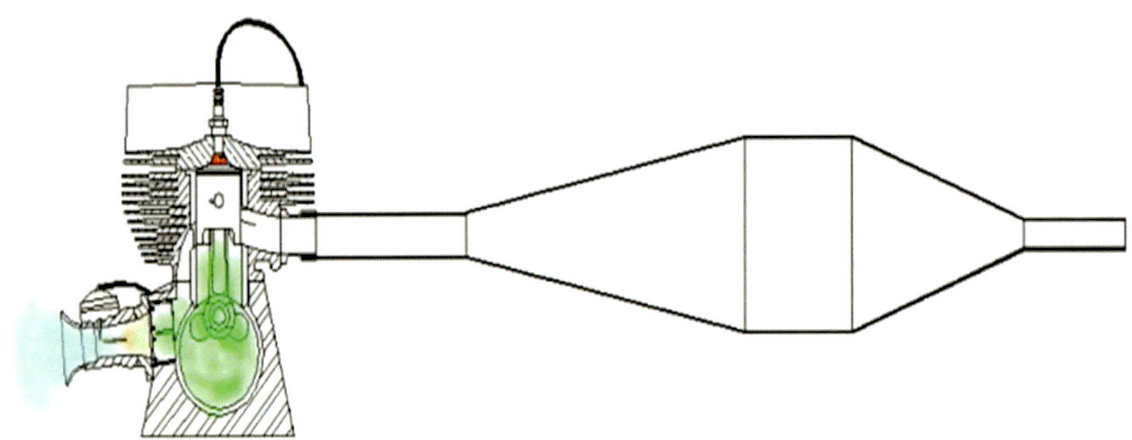

Two Stroke Engines

This is accomplished by using the end of the combustion stroke and the beginning of the compression stroke to perform simultaneously the intake and exhaust (or scavenging) functions

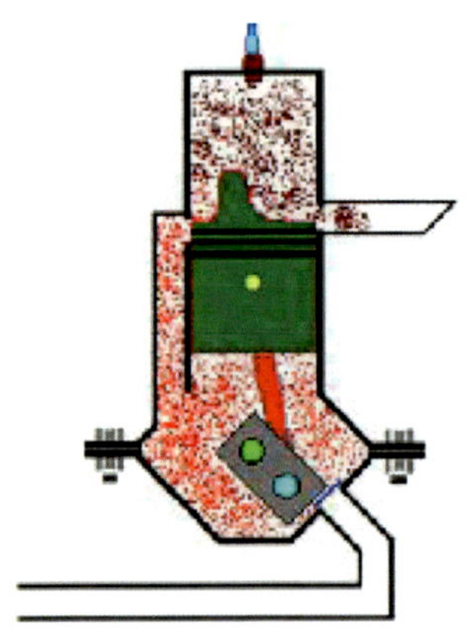

Scavenging Systems

scavenging is the **process** of pushing exhausted gas- charge out of the cylinder and drawing in a fresh draught of air or fuel/air mixture

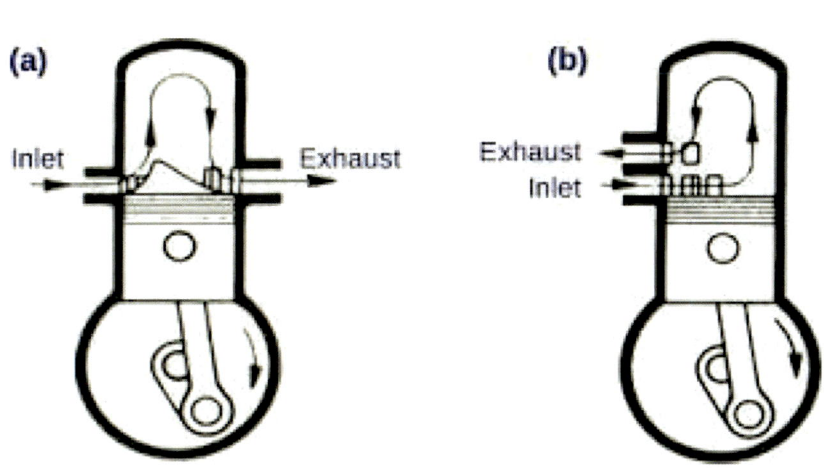

Scavenging process

- This process occurs after the exhaust port closes, which keeps the incoming fuel/air mixture from exiting the exhaust port.

- The fuel/air mixture that escapes during this process is what causes the issues that two-strokes have struggled with over the years

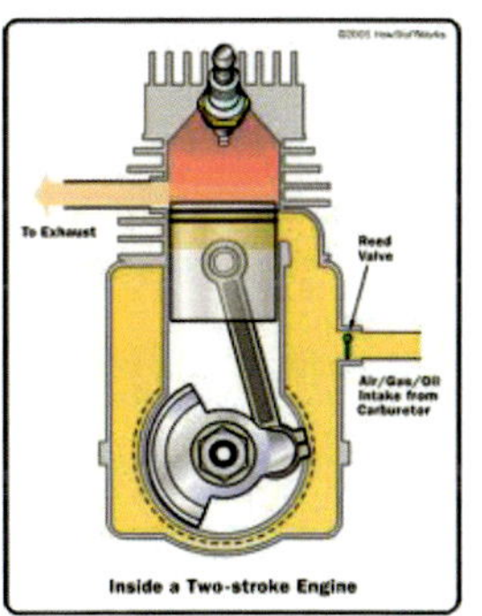

Two Stroke Gas Engines

No head valves. The cycle is complete in one crankshaft revolution

Piston Moving Up

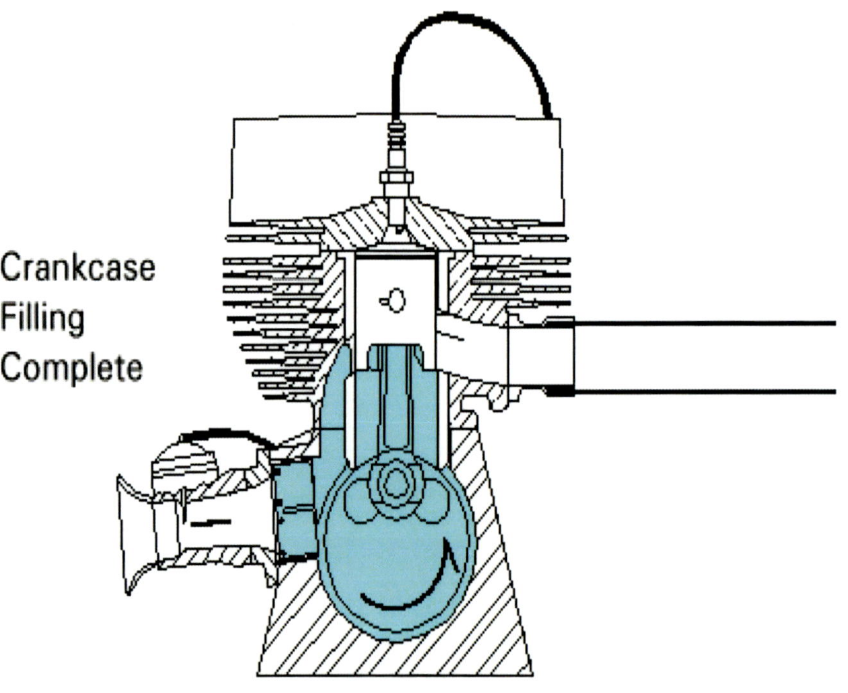

Piston Going Down

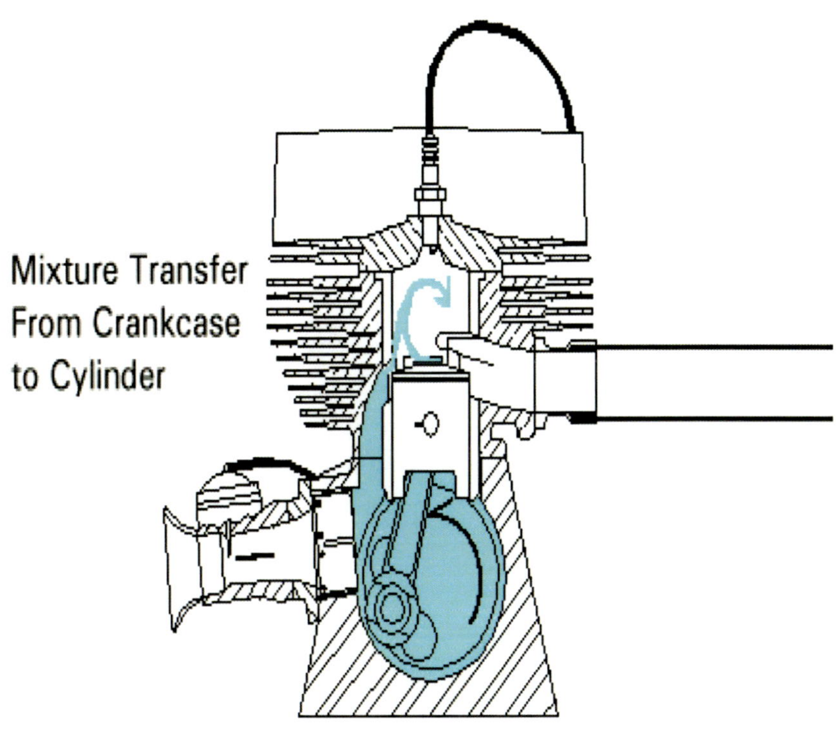

Compression Stroke

The trapped mixture is now compressed by the upward moving piston (at the same time that a new charge is being drawn into the crankcase down below)

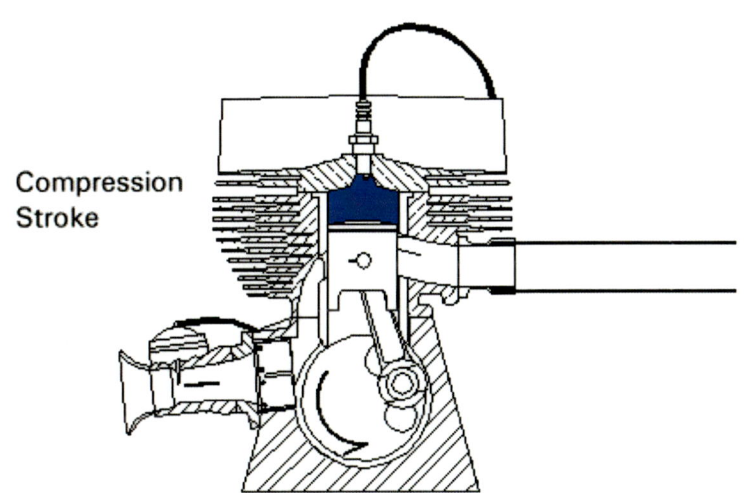

Exhaust Blow-down

The expanding mixture drives the piston downward until it begins to uncover the exhaust port. The majority of the pressure in the cylinder is released within a few degrees of crank rotation after the port begins to open

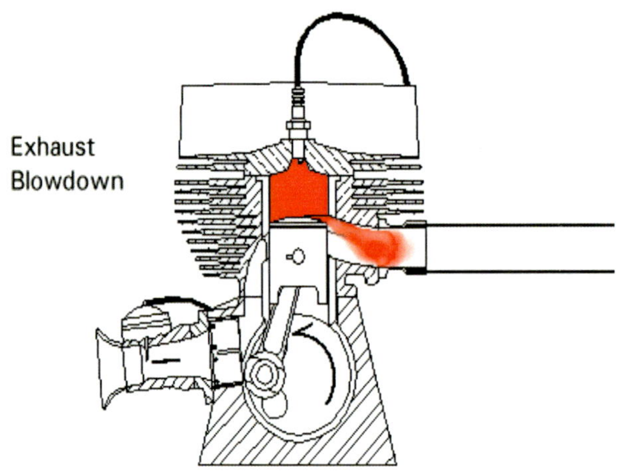

Two Stroke Pollution

Many designs use total-loss lubrication, with the oil being burned in the combustion chamber, causing "blue smoke" and other types of exhaust pollution. This is a major reason for two-stroke engines being replaced by four-stroke engines in many applications

Advantages of 2 Stroke Engines

- Two-stroke engines do not have valves, simplifying their construction

-

- 2 Stroke Head. 4 Stroke Head

Advantages of two strokes

- Two-stroke engines are lighter, and cost less to manufacture
- Two-stroke engines have the potential for about twice the power in the same size because there are twice as many power strokes per revolution

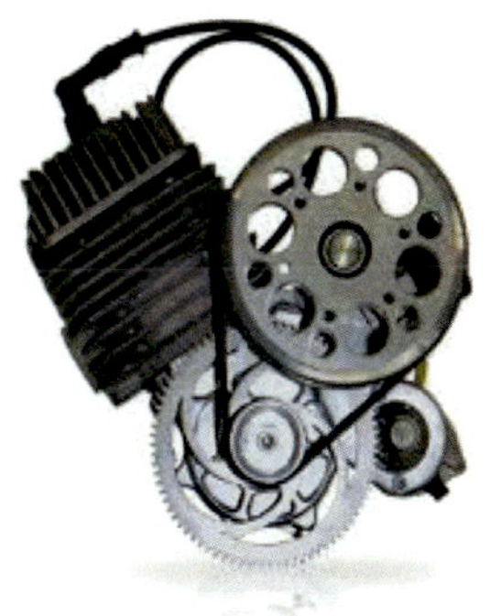

Disadvantages of two strokes

- Two-stroke engines don't live as long as four-stroke engines. The lack of a dedicated lubrication system means that the parts of a two-stroke engine wear-out faster.
- Two-stroke engines require a mix of oil in with the gas to lubricate the crankshaft, connecting rod and cylinder walls .

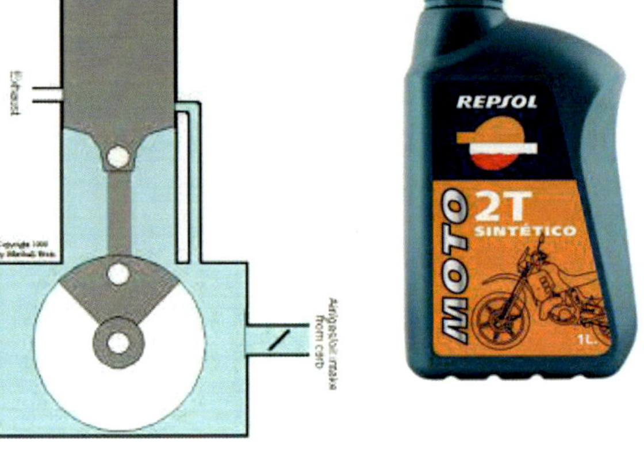

Pre Mix Gas & Oil

Many newer engines only require a 100:1 ratio, but 50:1 is the most common, and 24:1 is usually recommended during the break-in

Two-Stroke Pre-Mix Ratio Chart

Mixture Ratio to 1	Percent Oil	Ounces of Oil per Gallon of Gas					
		1 gal	2 gal	2.5 gal	3 gal	4 gal	5 gal
16	6.25	8.00	16.00	20.00	24.00	32.00	40.00
20	5.00	6.40	12.80	16.00	19.20	25.60	32.00
25	4.00	5.12	10.24	12.80	15.36	20.48	25.60
30	3.33	4.27	8.54	10.68	12.81	17.08	21.35
32	3.13	4.00	8.00	10.00	12.00	16.00	20.00
35	2.86	3.66	7.32	9.15	10.98	14.64	18.30
40	2.50	3.20	6.40	8.00	9.60	12.80	16.00
45	2.22	2.84	5.68	7.10	8.52	11.36	14.20
50	2.00	2.56	5.12	6.40	7.68	10.24	12.80
55	1.82	2.33	4.66	5.83	6.99	9.32	11.65
60	1.67	2.13	4.26	5.33	6.39	8.52	10.65
70	1.43	1.83	3.66	4.58	5.49	7.32	9.15
80	1.25	1.60	3.20	4.00	4.80	6.40	8.00
90	1.11	1.42	2.84	3.55	4.26	5.68	7.10
100	1.00	1.28	2.56	3.20	3.84	5.12	6.40

Cerma 2-Cycle Oils are produced using the best base oils and blended with STM-3 additive pack, to protect engine from wear, clean engine, increase usable power, and increase engine life.

50 to 1 Mixture

- A 50-to-1 mixture of gasoline and oil for two-cycle oil takes 50 parts gasoline to one part oil. This ratio can use any form of liquid measure; using 50 ounces of gasoline, for example, requires 1 ounce of oil to meet the criteria
- For Example in a 50 to 1 ratio, for 12 gallons of gasoline you should add 30.72 ounces of motor oil

MIX	RATIO	OIL	FUEL
Conventional	32:1	4.0 oz.	1 US GAL
Conventional	40:1	3.2 oz.	1 US GAL
Conventional	50:1	2.6 oz.	1 US GAL
SABER® Ratio™	80:1	1.6 oz.	1 US GAL
SABER® Ratio™	100:1	1.3 oz.	1 US GAL

Two Stroke Piston Rings

Two-stroke engines either come with one or two compression rings. One ring has less drag, revs better and makes it possible to produce a lighter piston. The disadvantage is that one ring can't provide as durable of a compression seal as two rings

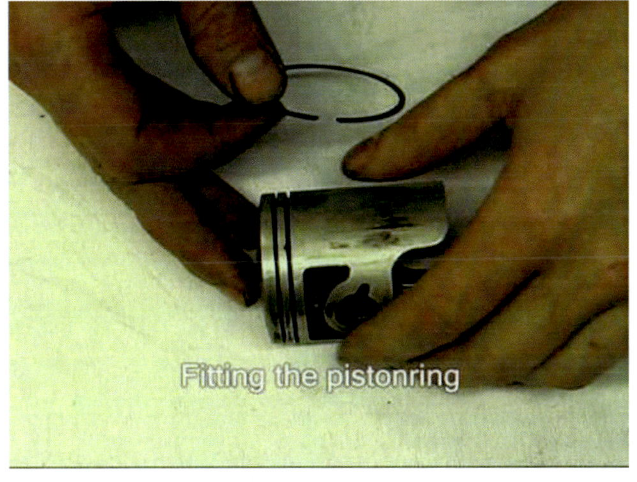

Piston Rings

- The two stroke engines only use two piston rings instead of three rings used in four stroke engines
- In four stroke engines the lower ring is for controlling the supply of oil to the liner which lubricates the piston skirt and the compression rings (oil control rings).

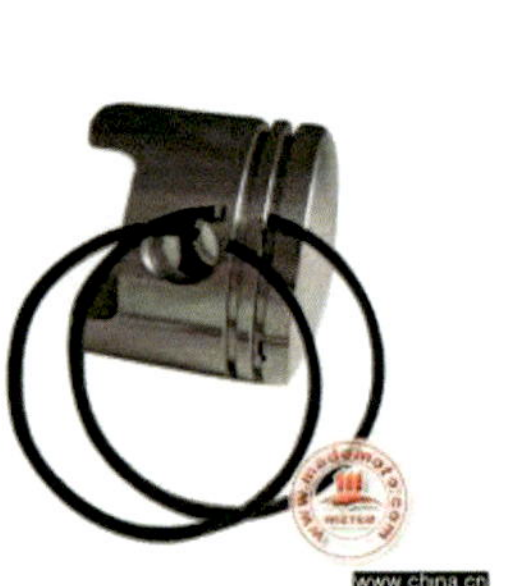

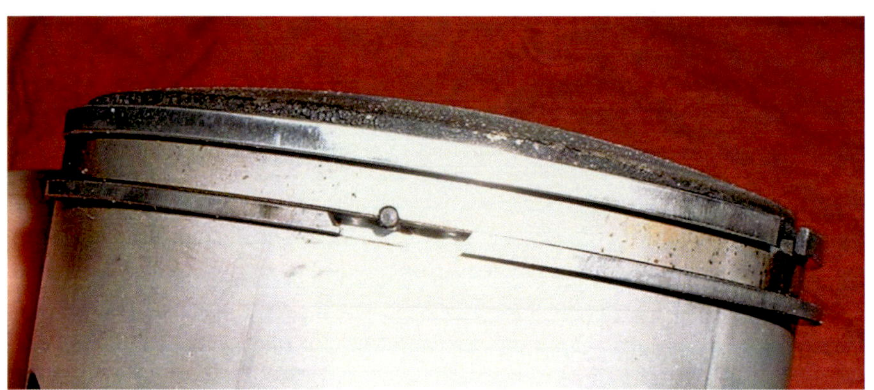

Piston Coatings

There are several different types of piston coatings, but the one that most people are familiar with is called a "break-in coating." This Teflon-like coating covers the skirt area of the piston to reduce friction and ease the break-in period.

Two Stroke Piston

- The top of the piston should be black. The sides of the piston crown should have little "pockets" on them

- You will be able to see the shiny aluminum in the pockets. These pockets are caused from cool gas flowing through the transfer ports and across that spot on the piston

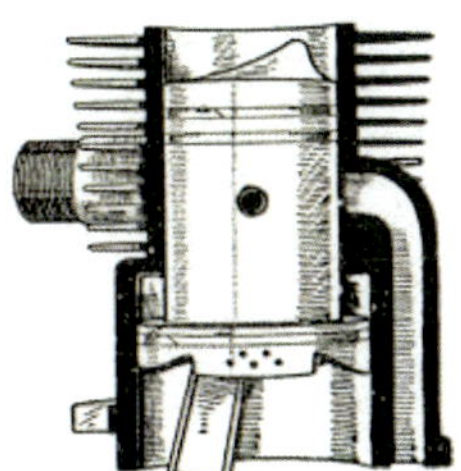 

PERFECT BROWN CROWN

The crown of this piston shows an ideal carbon pattern. The transfer ports of this two-stroke engine are flowing equally and the colour of the carbon pattern is chocolate brown. That indicates that this engine's carb is jetted correctly

BLACK SPOT HOT

The underside of this piston has a black spot. The black spot is a carbon deposit that resulted from pre-mix oil burning on to the piston because the piston's crown was too hot. The main reasons for this problem are overheating due to too lean carb jetting or coolant system failure

ASH TRASH

This piston crown has an ash color, which shows that the engine has run hot. The ash color is actually piston material that has started to flash (melt) and turned to tiny flakes

SMASHED DEBRIS

This piston crown has been damaged because debris entered the combustion chamber and was crushed between the piston and the cylinder head

SMASHED DEBRIS Causes

The common causes of this problem are broken needle bearings from the small or big end bearings of the connecting rod, broken ring ends, or a dislodged ring centering pin. When A problem like this occurs, its important to locate where the debris originated

CHIPPED CROWN DROWNED

This piston crown chipped at the top ring groove because of a head gasket leak. The coolant is drawn into the combustion chamber on the down-stroke of the piston

Examining a Used Piston

CHIPPED CROWN DROWNED: When the coolant hits the piston crown it makes the alumininum brittle and it eventually cracks. In extreme cases the head gasket leak can cause erosion at the top edge of the cylinder and the corresponding area of the head

SHATTERED

The skirts of this piston shattered because the piston to cylinder clearance was too great. When the piston is allowed to rattle in the cylinder bore, it develops stress cracks and eventually shatters

Engine Systems
(Inboards & Outboards)

In the next chapters we are going to analyze the following systems for both types of engines

- Fuel System

- Ignition System

- Cooling System

- Control and Diagnosis systems

- Lubrication System

- Exhaust System

Chapter 2
Fuel Systems
(Inboard & Outboards)

In-board Fuel Systems

The fuel systems for marine inboard gasoline engines are classified in :

- Carbureted engines or Naturally Aspirated
  - With electro-mechanical ignition system
  - With electronic ignition system and CPU
- Throttle body injection system (TBI) or single point fuel injection
  - With single or double injector at the throttle body plate
  - System fully controlled by a CPU
- Multi-port fuel injection (MPI) or Pre-combustion fuel injection system
  - Fuel Injectors located in the intake manifold just before the intake valves
- Direct fuel injection (DFI)
  - Fuel Injectors located on the head valves, directly into the combustion chambers

Carbureted System

The carburetor is located over the intake manifold and their function is to mix the Air and Fuel according with the stoichiometric ratio (14.7:1)

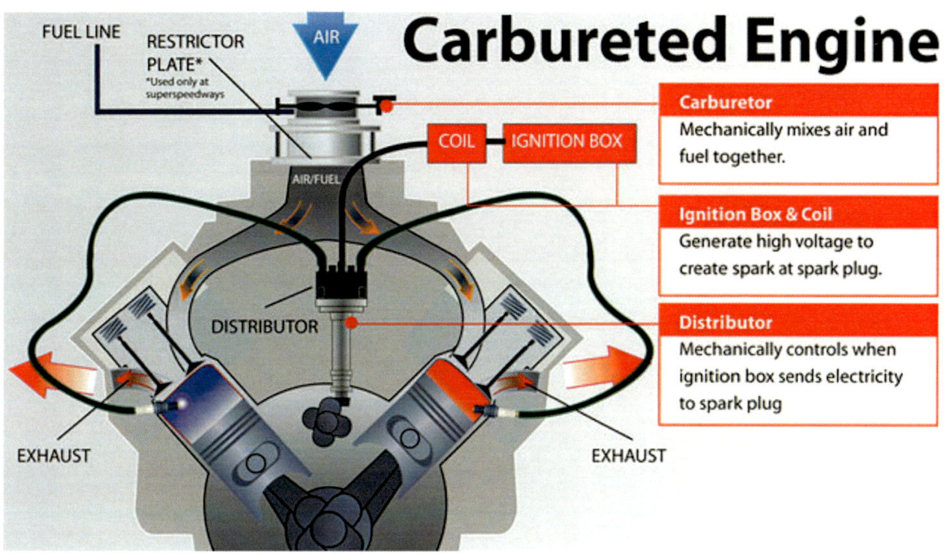

Fuel Pump / Transfer Pump

The fuel pump or transfer pump transport the fuel from the fuel tank into the carburetor, could be electrical or mechanical

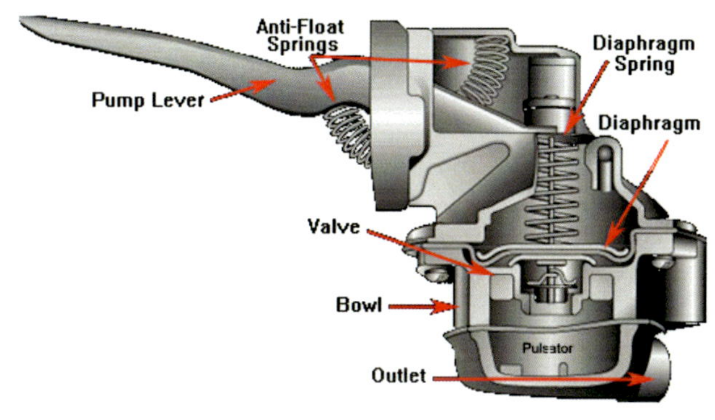

Fuel Pump / Transfer Pump

Is a positive displacement pump with a low output pressure , around 25 PSI

Fuel Pumps

The mechanical Pump is a diaphragm pump and the electrical pump is a rotary vane type

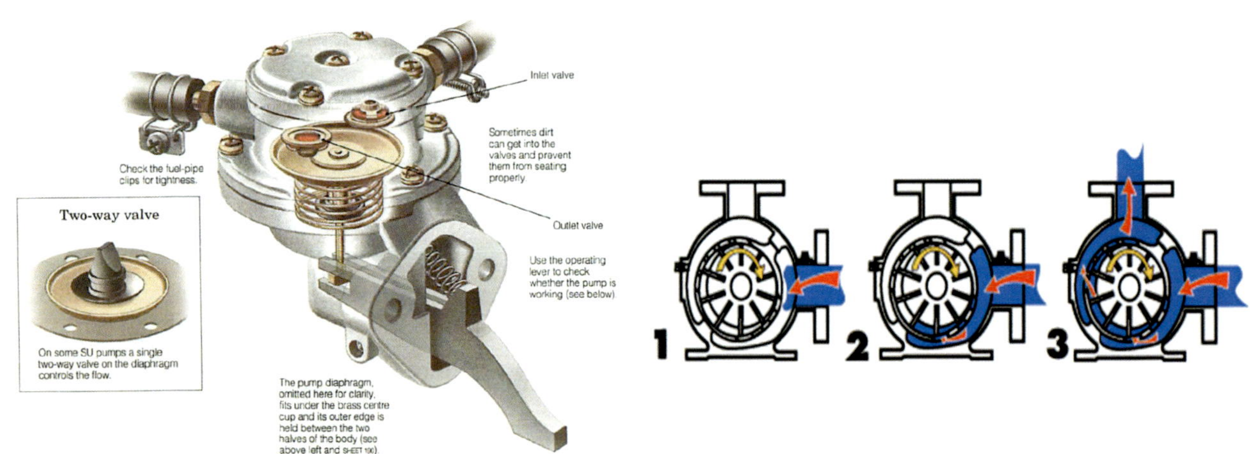

How to calculate the fuel pump

- To make this horsepower, your engine will consume a certain amount of fuel. That amount of fuel should be supplied through the fuel /transfer pump

- That amount is referred to as the "Brake Specific Fuel Consumption", or BSFC. The BSFC is generally estimated to be between 0.30 and 0.37 for most naturally-aspirated / Carbureted (non-turbo/super-charged) engines, and between .45 and .55 for turbo/super-charged engines

How much fuel your fuel pump needs?

- Multiply horsepower by .38 (naturally-aspirated motors) or .47 (force-induction motors) to come up with a fairly accurate guide to how many liters per hour of fuel you will need to feed the engine

- For example, if you are building a really hot little 4 cylinder turbocharged engine and plan to make about 450HP, you would need a fuel pump that can produce about about 212 liters per hour (450 * 0.47)

How much fuel your fuel pump needs?

- It is critical that the fuel pump in your fuel-injected engine is able to produce at least as much or more volume over time than the engine requires

- If the fuel pump is unable to meet the fuel requirements then the fuel mixture will become lean and the engine will go into pre-detonation and will eventually destroy itself

What is Bernoulli's principle?

So within a horizontal fluid pipe that changes diameter, <u>regions where the fluid is moving fast will be under less pressure than regions where the fluid is moving slow</u>.

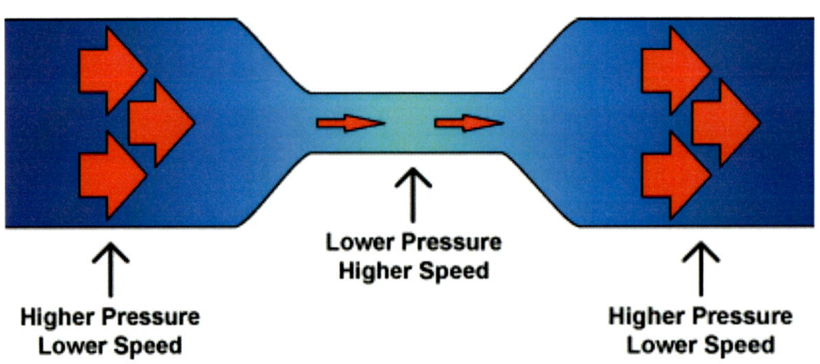

The Bernoulli Principle

This sounds counterintuitive to many people since people associate high speeds with high pressures. This is really just another way of saying that water will speed up if there's more pressure behind it than in front of it.

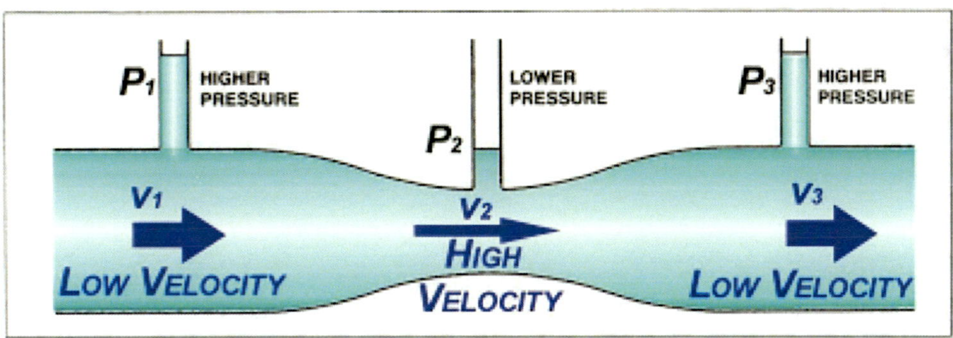

The Venturi Principle

A venturi is a streamlined restriction in any passage . This restriction causes an increase in air velocity and a lowering of pressure of the air passing through it. The greater the air velocity, the lower the air pressure at the venturi. This lower pressure is the basic force under which carburetors function.

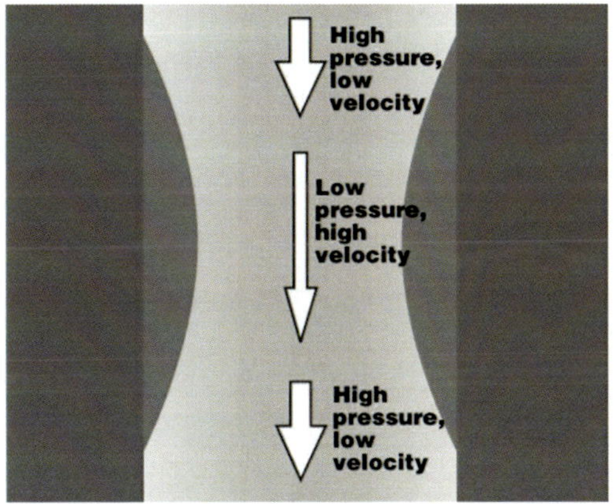

33

The Venturi Principle

The application of pressure difference is the basis of carburetor function. A venturi with a flap is incorporated in the throat or bore of all carburetors to establish this pressure difference

Velocity Vs. Pressure

The relationship of air velocity to air pressure can be illustrated by a simple experiment. Hold the edge of a sheet of paper to your lower lip, allowing the rest of the sheet to hang limp. By blowing across the top of the paper, you will notice the paper rises. Air in motion across the top of the paper exerts less pressure than the normal atmospheric pressure of the stationary air under the paper

The Choke Valve

The choke, which can be either manual or electrically operated, is a small flap that precedes the venturi tunnel . When a motor is cold, it is hard to keep running, so the choke door closes, thereby cutting down airflow and drastically richening up the air/fuel ratio

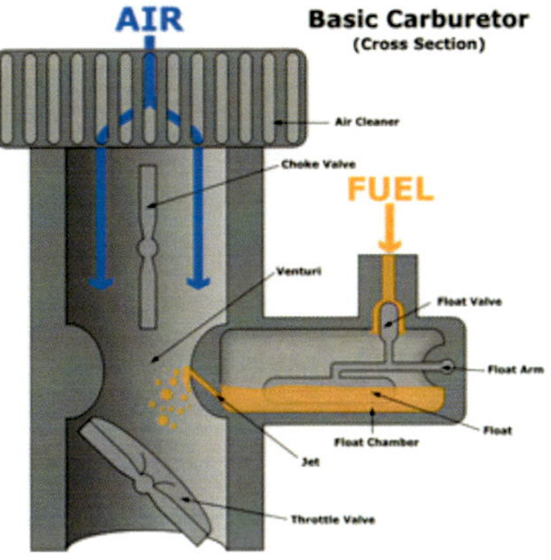

The Venturi Tunnel

The mixture, air and gas is mixed inside of the carburetor specifically through the venturi tunnel , where the differential pressure (Bernoulli Principle) , suctions the air and the gasoline at high speed

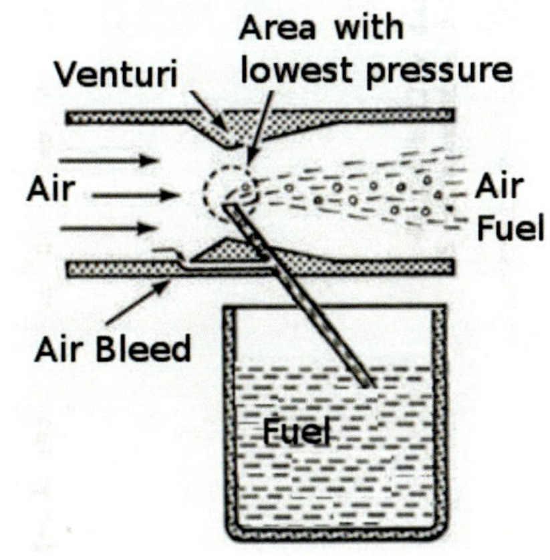

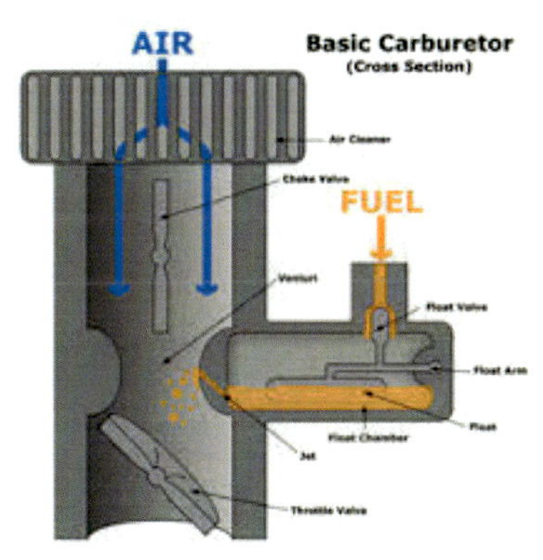

Two Barrels Carburetors

There are carburetors with one, two or four barrels and one venturi tunnel per each barrel

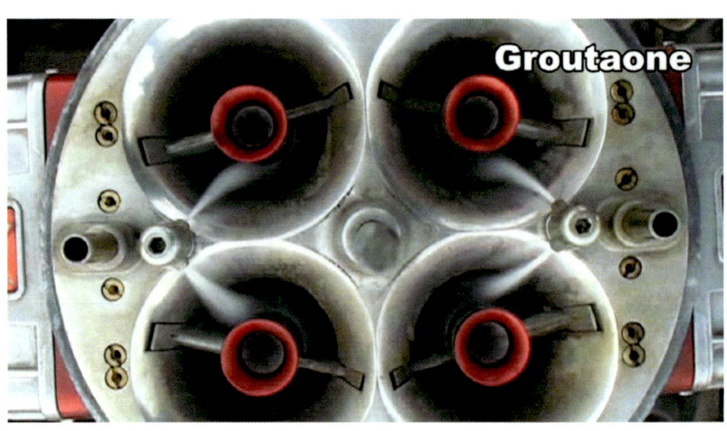

Four Barrel Carburetor

The four barrel carburetor has a sequential mechanism. In other words, the first two barrels open from idle and just before the peak torque point ; after that point the flaps of the other two barrels start to open

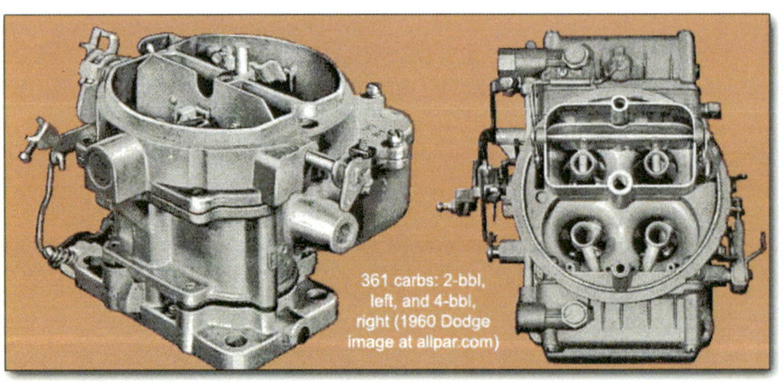

Floats and Float Level

The float ball is a small plastic or metal cylinder that sits inside the fuel reservoir . The float bowl with the seat and needle, determines the amount of gas flowing into the carburetor

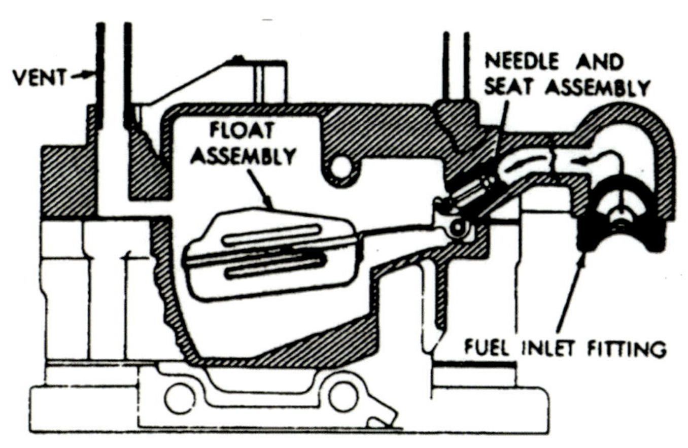

Ball , Needle and Seat

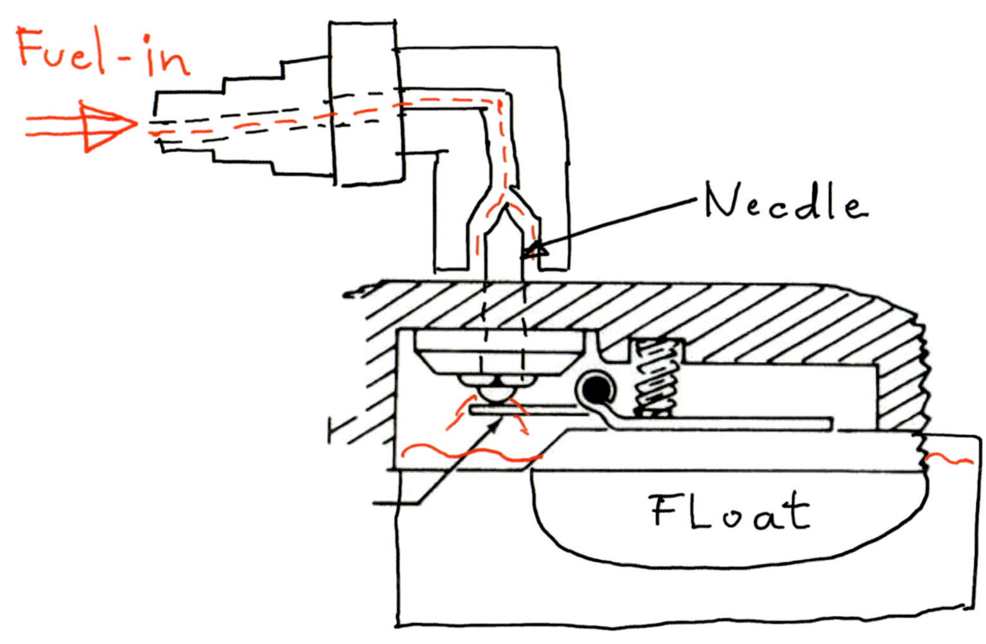

Idle Circuit

Since the vacuum created by air's rushing through the venturi at speed is what normally draws fuel into the engine while the motor is at speed, a different solution had to be thought up for when the motor is idling and the throttle blades are barely open

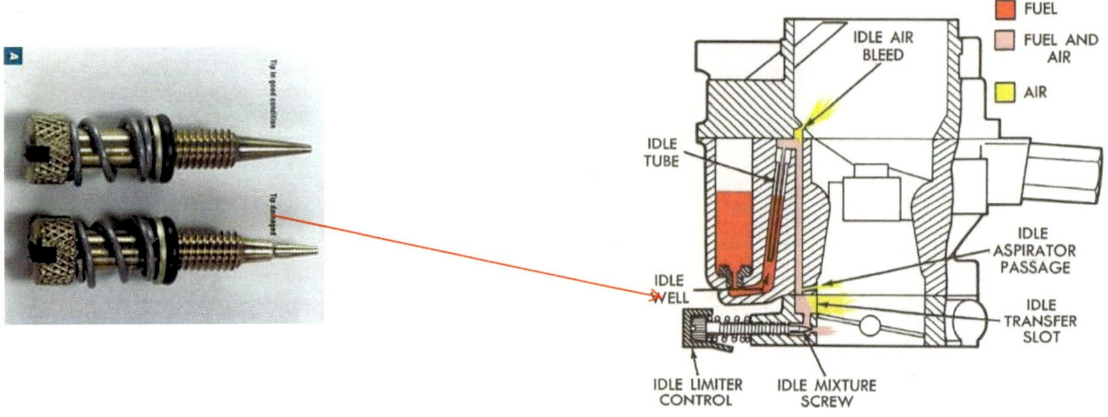

Idle Circuit

The idle circuit supplies fuel in this situation, and mixture screws let you adjust the air/fuel mixture of this circuit.

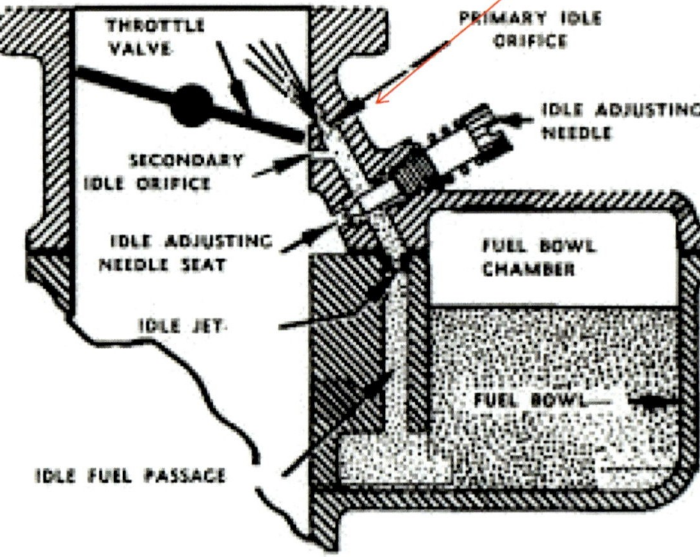

JETS

The jets are small, threaded plugs in the base of
the carburetor in the metering block which regulate
the amount of fuel that flows from the float bowl
into the venturi

Jets Diagnosis

The easiest way to check your jetting without an
air/fuel meter is   accelerating hard, then shut the
motor off before it has a chance to idle.Then remove
a few spark plugs and check the color of the
porcelain. It should be a nice light-brown-cocoa color.

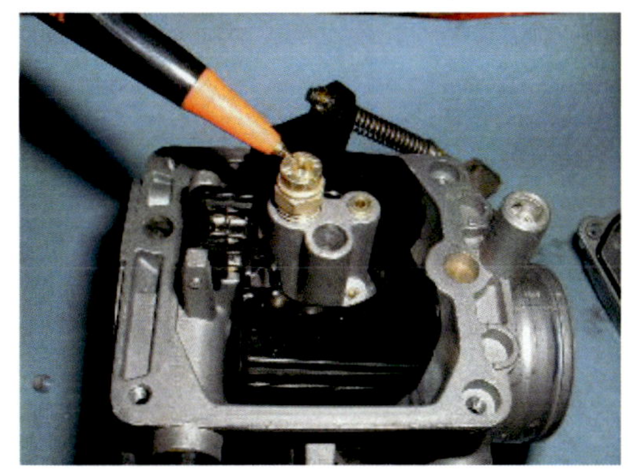

Jets Diagnosis

If it's any lighter, your motor is running lean, which could hurt power and, in a worst-case scenario, even burn a piston. If the motor is running rich, the color could be dark brown or even black. A good way to adjust jet size is to go up or down as necessary two sizes at a time until the optimal ratio is set.

Outboard Carburetors

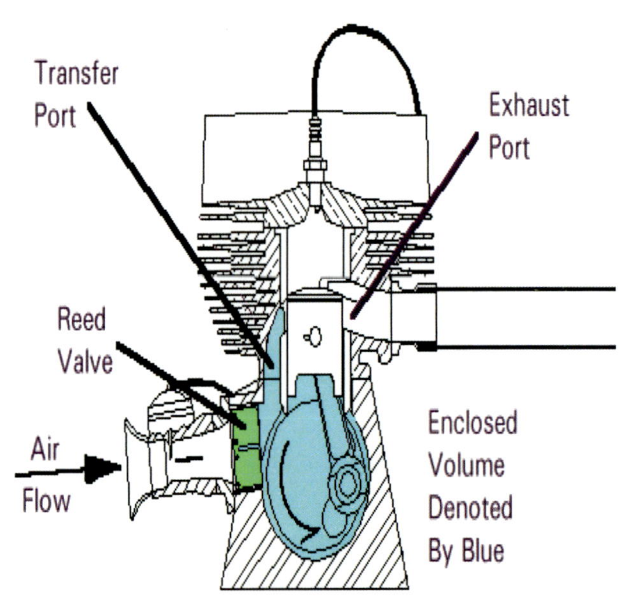

Outboard Carburetor

- Single-cylinder, and two or three-cylinder outboards will normally employ just one carburetor – a single-barrel float-equipped model that possesses either
- *An un-adjustable high-speed* jet and an adjustable low or idle-speed jet
- *un-adjustable high and low*-speed jets
- *adjustable high and* low-speed jets

Two Barrel Carburetor

Bigger outboard motors instead harbor one or more two-barrel carburetors – each which has two jets, a high-speed jet and a low or idle-speed jet, per barrel

Lean and Rich Mixture

If you wish to lean out the fuel-air mixture: or decrease the amount of fuel relative to air, then you need to *reduce* the jet size; focus on the medium and high-speed jets

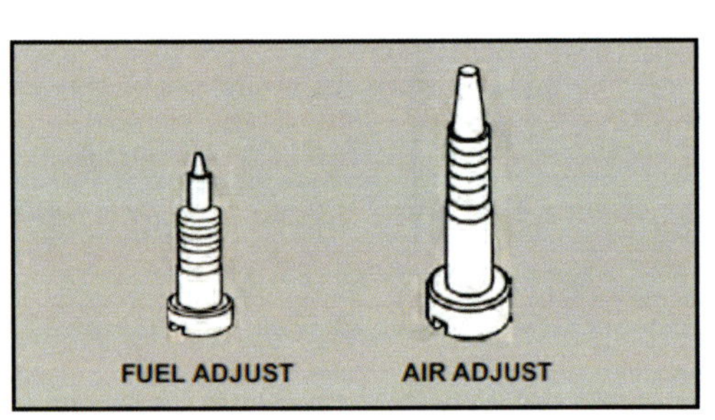

Lean and Rich Mixture

To enrich the fuel-air mixture: then *increase* the jet size, focusing your attention, again, on the medium and high-speed jets

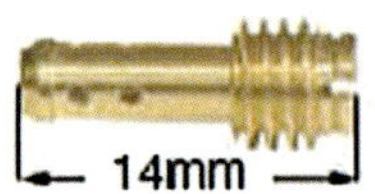
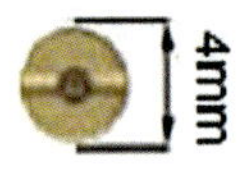

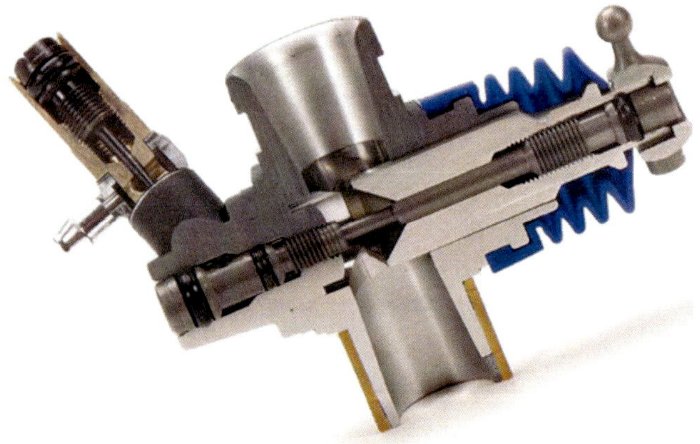

Air Bleed Jets

An air-bleed jet controls the amount of *air* in a fuel-air mixture, and therefore functions opposite to the medium and high-speed jets

Air Bleed Jets

Reducing an air-bleed jet size will restrict air relative to fuel intake and actually enrich a fuel-air mixture; increasing the size of an air-bleed jet will usher in more air, and make a mixture lean

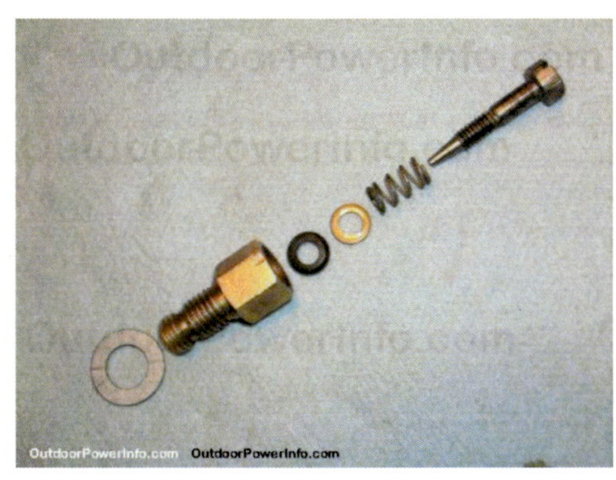

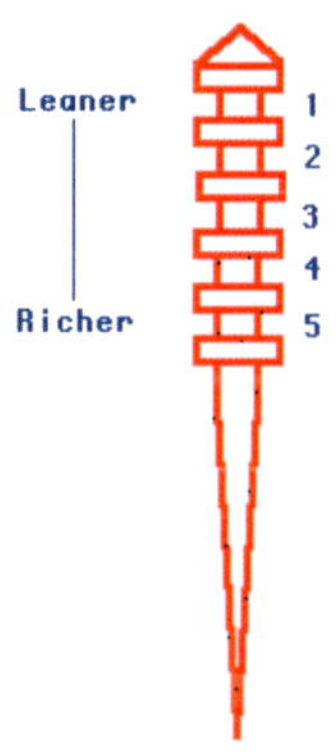

Outboard Carburetor Adjustment

- Before an outboard carburetor is adjusted, the engine should be at normal operating temperature. On most outboard carburetors, only the low-speed jet can be adjusted; the high-speed jet has fixed internal dimensions that are designed to offer smooth operation over a wide array of conditions
- Check the engine operational manual in order to identify where are located the high and low speed adjustable jets and the air bleed jets

High Speed Jet adjustment

- To adjust the high speed jet:
  - *1)open the engine throttle* to "full"
  - *2)if you have a tachometer to measure engine rpm* during adjustment of the jet, so much the better.
- Note:Any carburetor adjustment, the motor should be given 20 or so seconds to adapt to the new jet setting before its suitability is judged. Accelerate and decelerate the engine to check the engine response

Low Speed Jet Adjustment

- The motor should be operated at roughly 750 rpm. The goal for setting a low-speed jet is as noted above, the maximum amount of engine smoothness obtainable without drastically affecting power or rpm

- Note: if a carburetor contains two adjustable jets, it is a good idea to tweak the high-speed jet again after the low-speed jet has been adjusted!

Power Reed Valves

A reed valve consists of flexible reed petals that sit over an opening in a wedge-shaped block. The block fits between the carburetor and the engine.

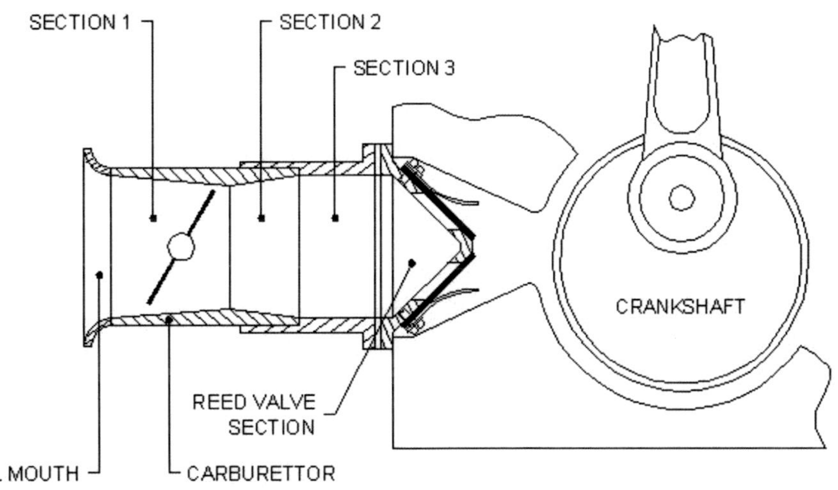

Power Reed Valves

With a lower pressure zone on the engine side, the reed petals flex open to allow fuel/air to pass, when the lower pressure switches to the intake tract side, the reed petals are forced tight against the reed block to seal off the intake tract

Red Valves / Flapper valves

Red Valves control the fuel-air mixture admitted to the cylinder. As the piston rises in the cylinder a vacuum is created in the crankcase beneath the piston

Red Valves / Flapper valves

This vacuum opens the valve and admits the fuel-air mixture into the crankcase. As the piston descends, it raises the crankcase pressure causing the valve to close to retain the mixture and pressurize it for its eventual transfer through to the combustion Chamber

Power Reed Valves

The reed petals pulse as the engine cycles, at roughly a one-to-one ratio. When the engine turns 8000 engine revolutions per minute, a reed opens 7980 times per minute. Needless to say, when the engine is running close to peak rpm, the reeds are really buzzing

Throttle Body Injection (TBI)

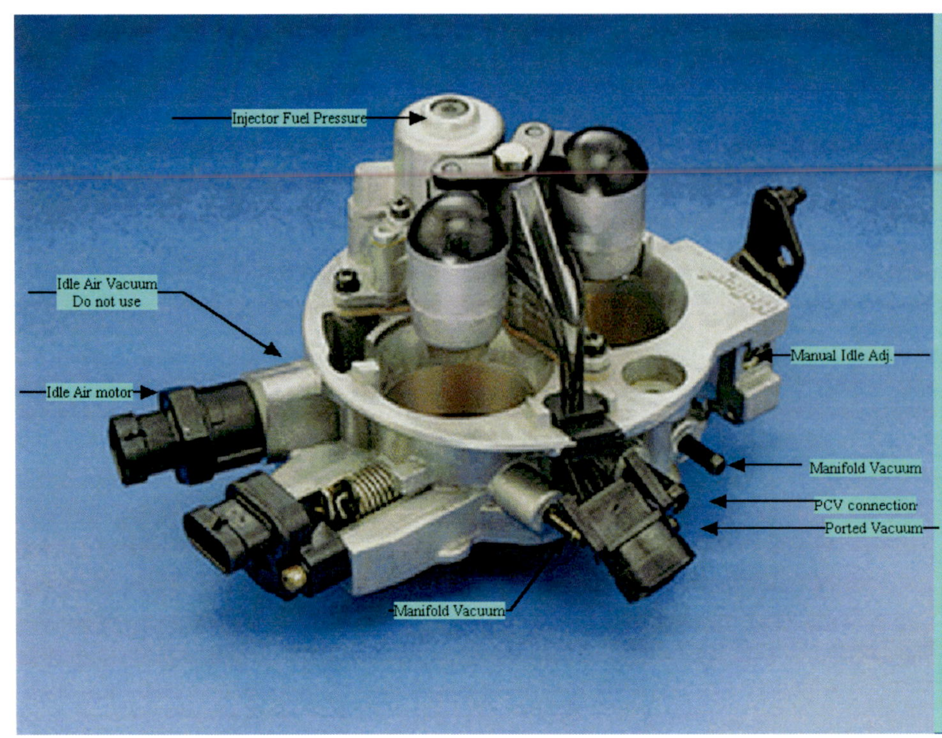

Throttle Body Injection (TBI)

This type of fuel injection system consists of only two major castings - **the fuel body** and the **throttle body**

TBI Fuel Injection Advantages

- It is less expensive than using other types of fuel injection systems

- It is easier to clean, maintain and service because there are fewer parts

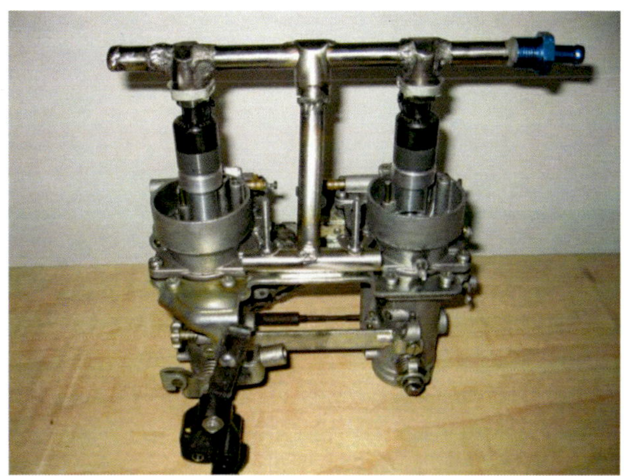

TBI Fuel Injection Advantages

- It is cheaper to manufacture than a port injection system and simpler to diagnose. It also does not have the same level of injector balance problems that a port injection system might have when the injectors are clogged

- It greatly improves the fuel metering compared to a carburetor

TBI Fuel Injection Disadvantages

It is almost the same as a TBI carburetor wherein the fuel is not equally distributed to all the cylinders. This means that the air/fuel mixture injected differs for each cylinder

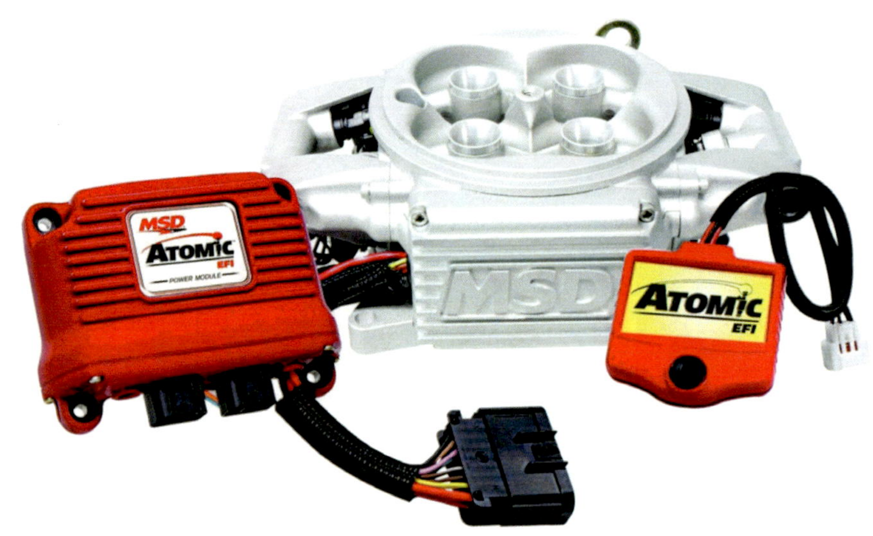

TBI Fuel Injection Disadvantages

- It can cool the manifold much faster causing the fuel to puddle and condense in the manifold. The possibility of condensation is much higher since the fuel travels longer from the throttle body to the combustion chamber

TBI Fuel Injection Disadvantages

Since the system needs to be mounted on top of the combustion chamber, you're prevented from modifying the manifold design to improve your car's performance

Multi Port Fuel Injection MPFI

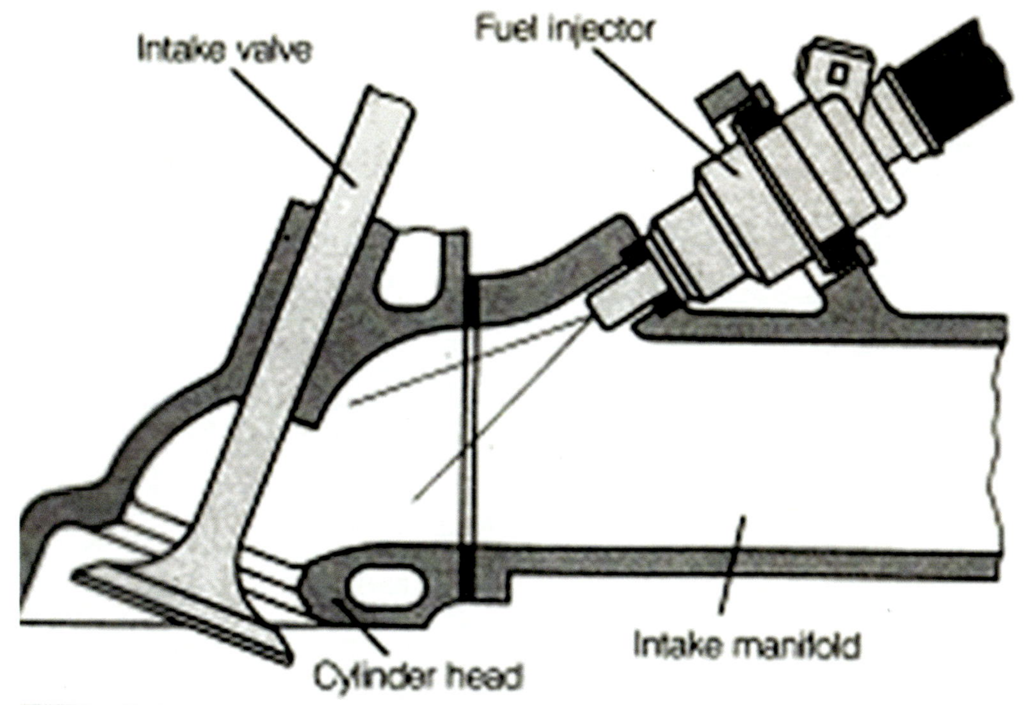

Multi Port Fuel Injection

Multi port fuel injection injects fuel into the intake ports of each cylinder's intake valve, rather than at a central point within an intake manifold like in spark plugs. It can be sequential, in which injection is timed to coincide with each cylinder's intake stroke

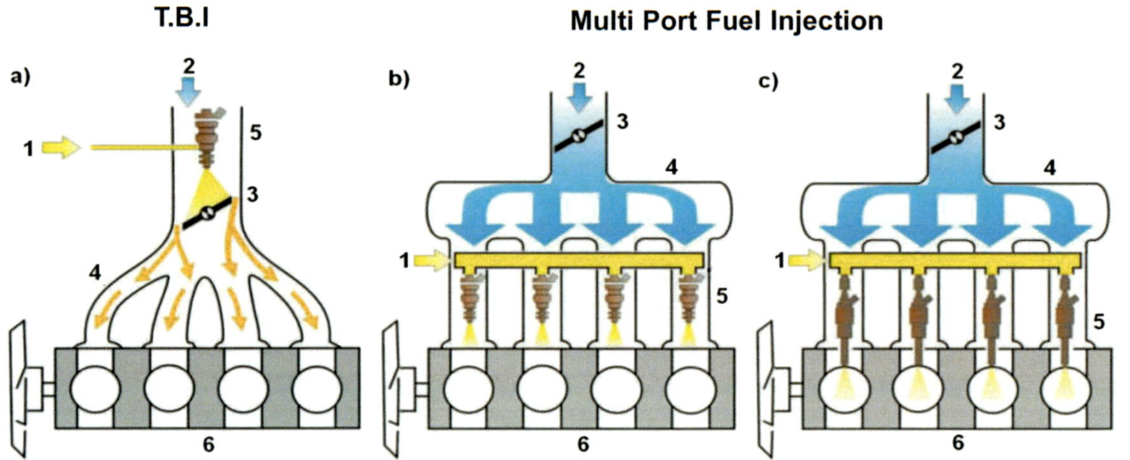

Multi Port Fuel Injection

When air flows through the intake manifold, the intake manifold remains dry and there are no problems with fuel puddling or fuel separation causing uneven fuel mixtures in the center and end cylinders

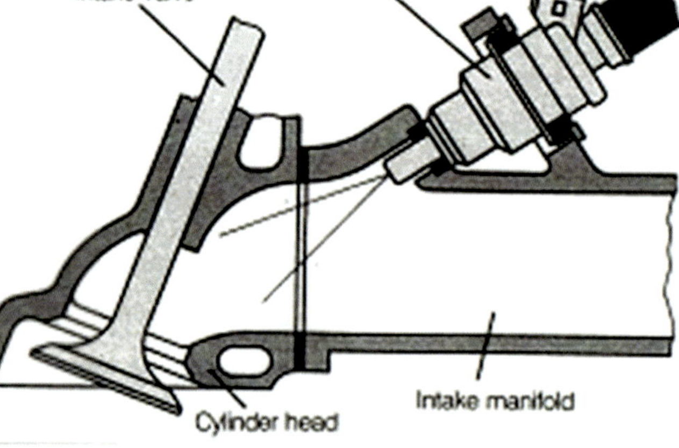

Sequential Fuel Injection

Sequential fuel injection is a type of multi-port fuel injection system in which each injection valve will open just before the cylinder intake valve opens. So in essence, the individual injectors work by themselves because they are fired individually.

In other words "Sequential Fuel Injection" and "Multi Port Fuel Injection" are basically the same concepts

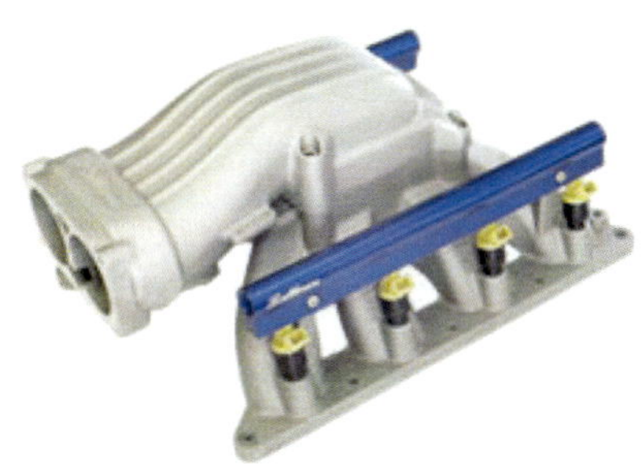

Direct Fuel Injection

The gasoline is highly pressurized, and injected via a common rail fuel line directly into the combustion chamber of each cylinder

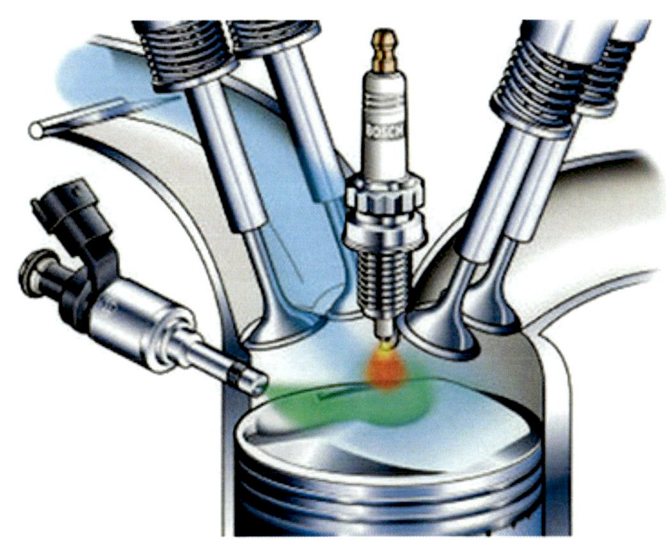

Direct Fuel Injection

Directly injecting fuel into the combustion chamber requires high pressure injection whereas low pressure is used injecting into the intake tract or cylinder port

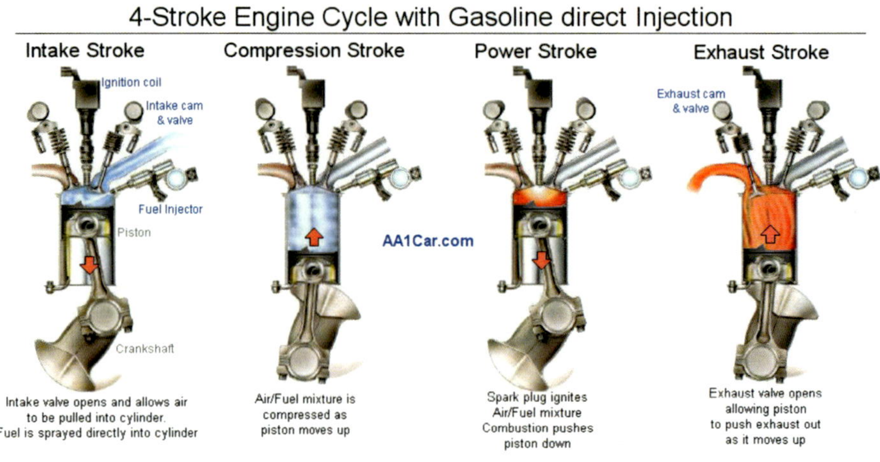

Direct Fuel Injection

By putting the injector inside the cylinder, the engine's computer gains even more precision control over the amount of fuel during the intake stroke, further optimizing the air/fuel mixture to create a clean burning explosion with very little wasted fuel and increased power delivery

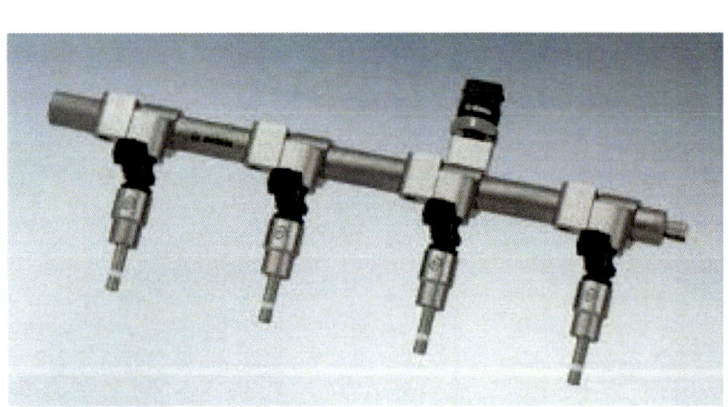

Gasoline Direct Injection

GDI engines can adjust the timing so that a smaller amount of fuel is injected during the compression stroke, creating a much smaller, controlled explosion in the cylinder. This so-called ultra lean burn mode sacrifices a bit of outright power, but greatly reduces the amount of fuel used during times when the vehicle requires very little grunt (idling, coasting, decelerating, etc.)

Pressure on DGI Systems

The injectors on a GDI system must be more rugged than port injectors in order to withstand the heat and pressure of hundreds (or even thousands) of tiny explosions per minute

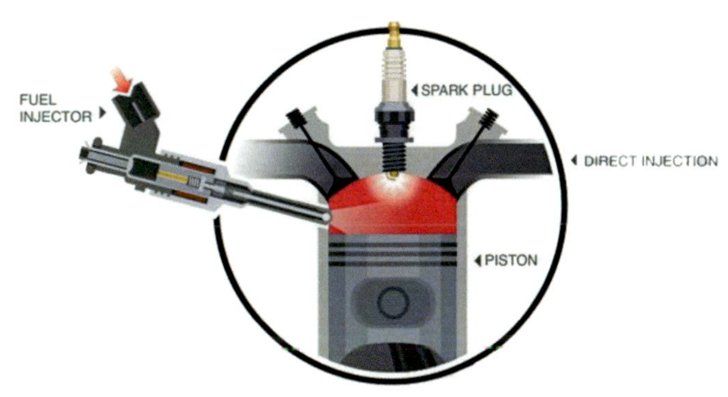

Pressure on DGI Systems

A GDI system needs to be able to inject fuel into a pressurized combustion chamber, the fuel lines supplying the gasoline need to be even higher in compression. GDI fuel systems can run at many thousand psi versus the 40 to 60 psi of port injection systems

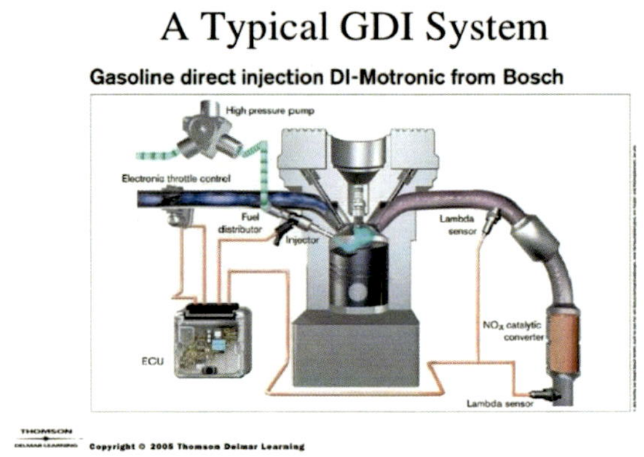

A Typical GDI System

GDI & Turbocharger

Owners of GDI engines (particularly higher-performance, turbocharged models) have reported that direct-injection systems see increased carbon buildup in backsides of their intake valves, resulting in reduced airflow and performance over time

Gasoline Direct Injection Engine Valves

Intake Valve Carbon Deposits

After Cleaning

GDI Carbon Deposits

The buildup occurs because in most engines intake air is, frankly, kind of dirty -- even with air filters in place, modern exhaust gas recirculation systems and crankcase vent systems can add quite a bit of muck to the intake charge

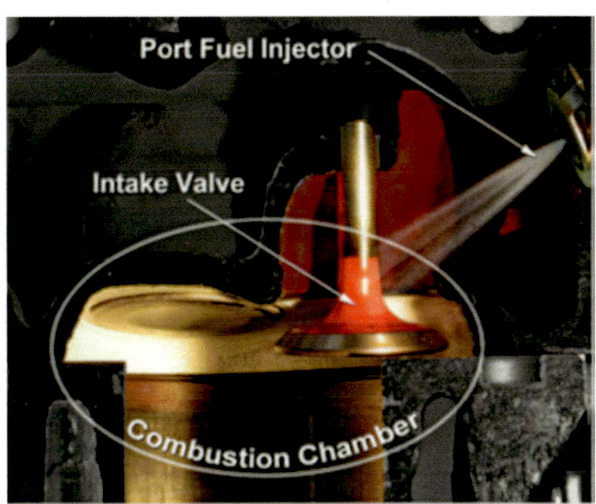

Engine Control Unit - ECU

The ECU is primarily responsible for four tasks. Firstly, the ECU controls the fuel mixture. Secondly, the ECU controls idle speed. Thirdly, the ECU is responsible for ignition timing. Lastly, in some applications, the ECU controls valve timing

The Gasoline Path

The gasoline on the fuel tank gets sucked up by an electric fuel pump. In some cases the electric pump or transfer pump is located between the fuel tank and the engine in other cases the fuel pump usually comes in an in-tank module that consists of a pump, a filter, and a sending unit

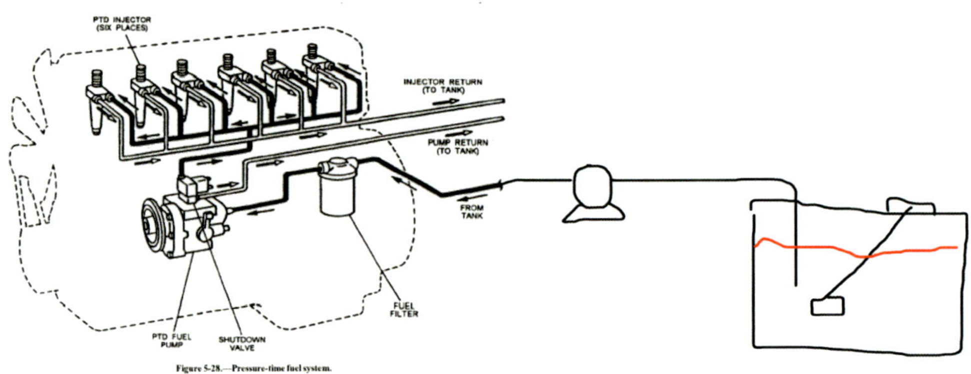

The Gasoline Path

The pump sends the gasoline through a fuel filter, through hard fuel lines, and into a fuel rail

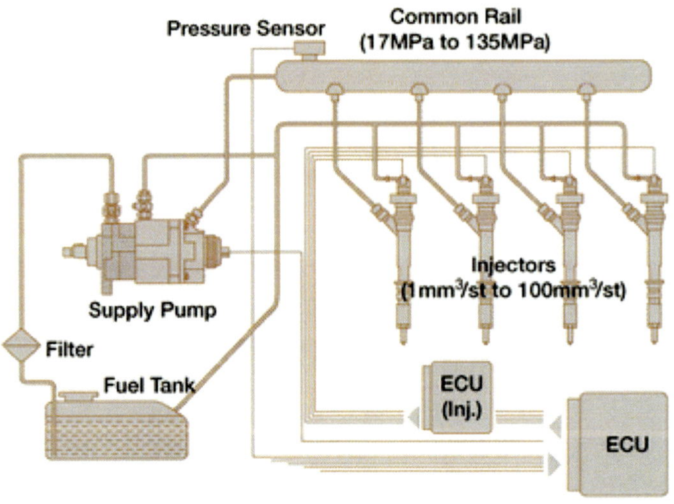

The Gasoline Path

A vacuum-powered fuel pressure regulator/valve at the end of the fuel rail ensures that the fuel pressure (35 to 50 PSI) in the rail remains constant relative to the intake pressure

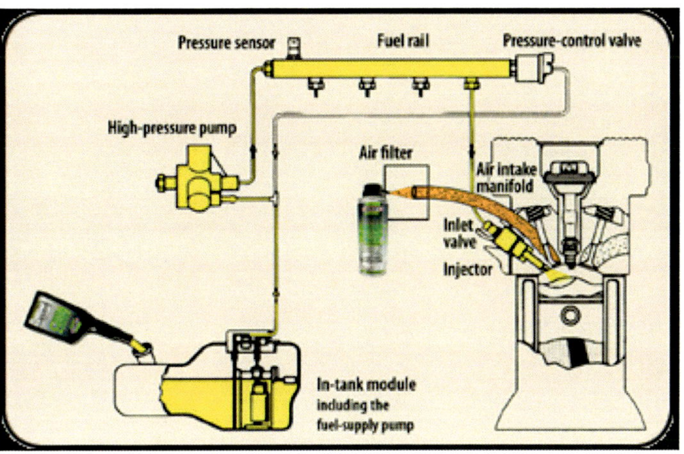

The Gasoline Path

Fuel injectors connect to the rail, but their valves remain closed until the ECU decides to send fuel into the cylinders.

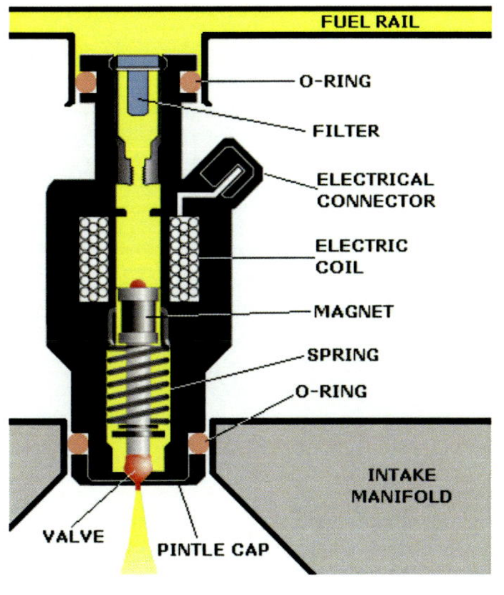

Fuel Injector Wiring

Usually, the injectors have two pins. One pin is connected to the battery through the ignition relay and the other pin goes to the ECU. The ECU sends a pulsing ground to the injector, which closes the circuit, providing the injector's solenoid with current

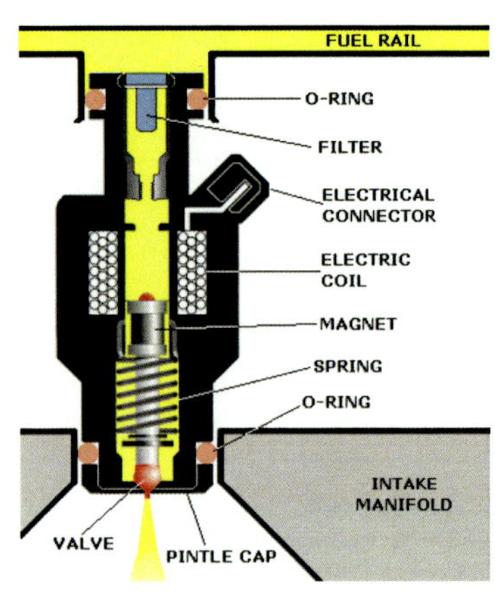

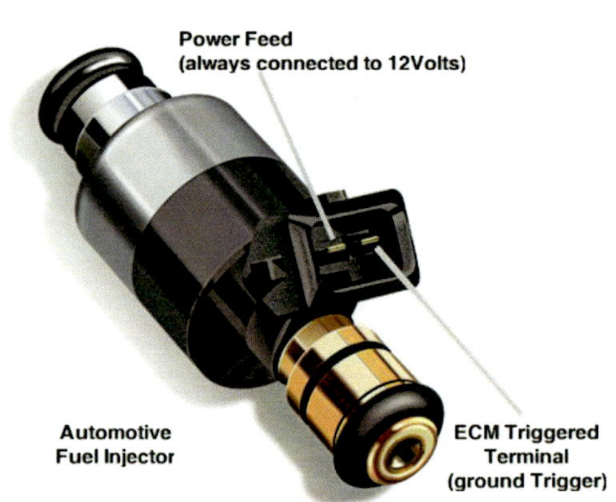

Fuel Injector Wiring

The magnet on top of the plunger is attracted to the solenoid's magnetic field, opening the valve. Since there is high pressure in the rail, opening the valve sends fuel at a high velocity through the injector's spray tip

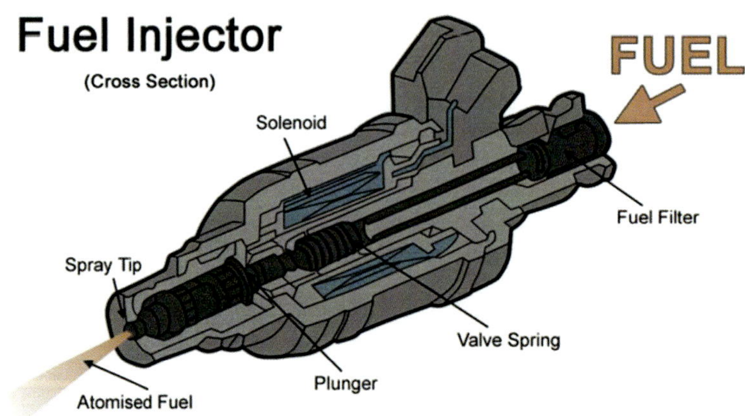

Cleaning Fuel Injectors

If the fuel injectors are restricted or leaking, there are several options to clean them including:

- **A fuel injector-cleaning chemical that is poured into the fuel tank.**
  Provides a noticeable difference in power and drive-ability after one use. The disadvantages are that you can't check the spray pattern

GDI Before Cleaning   GDI After Cleaning

Cleaning Fuel Injectors

If the fuel injectors are restricted or leaking, there are several options to clean them including:

- **An ultrasonic injector cleaner and injector flow bench.**
  The preferred method of cleaning injectors. Cycles the fuel injectors while the fuel injector tips are immersed in an ultrasonic bath, which uses an environmentally friendly cleaning fluid to clean the fuel injectors

Cleaning Fuel Injectors

The flow bench can simulate any rpm or load condition while the technician can observe the spray pattern. Ideally, all fuel injectors would be built to the same flow tolerances

Compression & Vacuum Test

- Engine compression is most accurately checked with a compression gauge. This test can take considerable time, especially when the spark plugs are difficult to remove

- A quicker test is a manifold vacuum test. If you have over 3" of cranking vacuum, there should be enough compression to run the engine

Compression Loss

- The first indication of the blown head gasket is an audible hissing or tapping sound, which can be difficult to locate, as it sounds like a bad lifter or rocker arm

- Depending on the location, you may see the air leak on the engine. The most accurate test for this problem is a compression gauge

Compression Tester

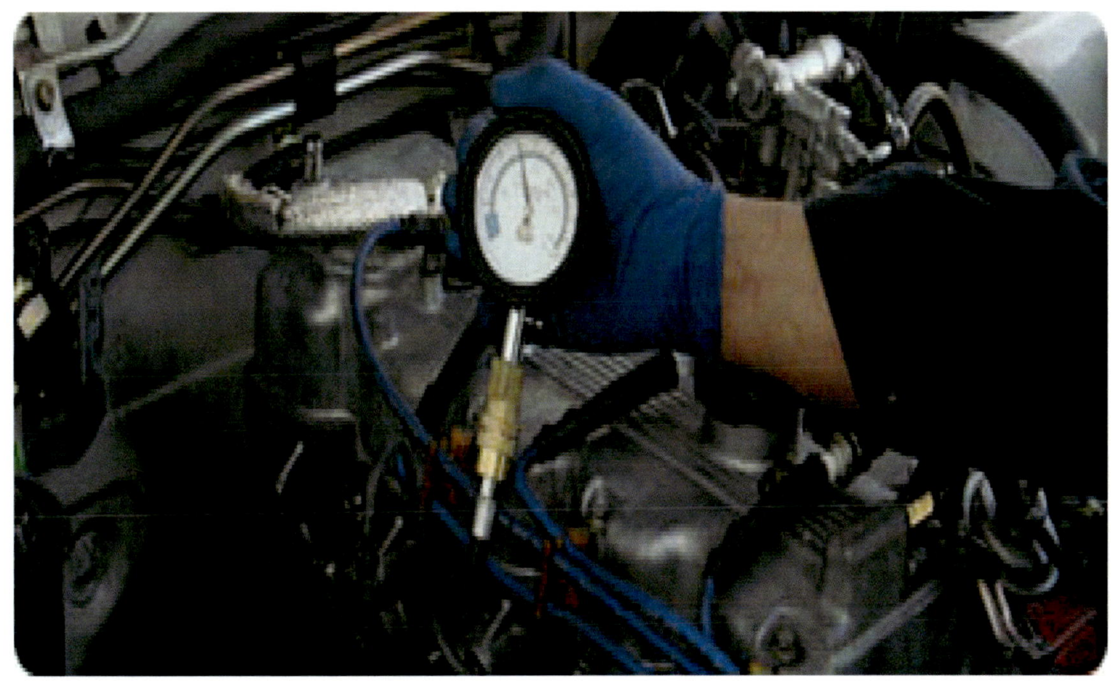

Engine Compression Testing

- An engine is essentially a self-powered air pump , so it needs good compression to run efficiently, cleanly and to start easily
- As a rule, most Gas engines should have 140 to 160 lbs. Of cranking compression with no more than 10% difference between any of the cylinders
- For Diesel engines the cranking compression are between 400 and 1200 PSI

HOW TO CHECK COMPRESSION

- All the spark plugs must be removed
- The ignition coil must then disabled or the high tension lead grounded
- If the engine has a distributorless ignition, the ignition coils must be disabled to prevent them from firing
- The throttle must also be held open.

Low Compression Evaluation

- Low compression in one cylinder usually indicates a bad <u>exhaust valve</u>
- Low compression in two adjacent cylinders typically means you have a bad <u>head gasket</u>

Low Compression??

Low compression in all cylinders would tell you That the rings, valves, gaskets and cylinders are worn and the engine needs to be overhauled

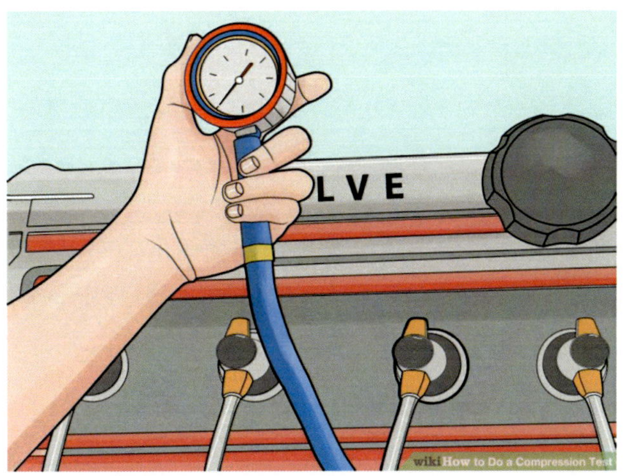

Piston Rings or Valves

- If compression is low in one or more cylinders, you can isolate the problem to the valves or rings by squirting a little 30 weight motor oil into the cylinder through the spark plug hole and repeating the compression test. The oil temporarily seals the rings.

- If the compression readings are higher the second time around, it means the rings and/or cylinder is worn

- No change in the compression readings would tell you the cylinder has a bad valve

Vacuum Test

- There are three types of engine vacuum in gasoline engines. They are Manifold Vacuum, Ported Vacuum, and Venturi Vacuum

- The most common is the manifold vacuum. This is the vacuum created between the throttle plate and the intake valves. Normal manifold vacuum on an engine at idle speed is around 18 to 20 inches

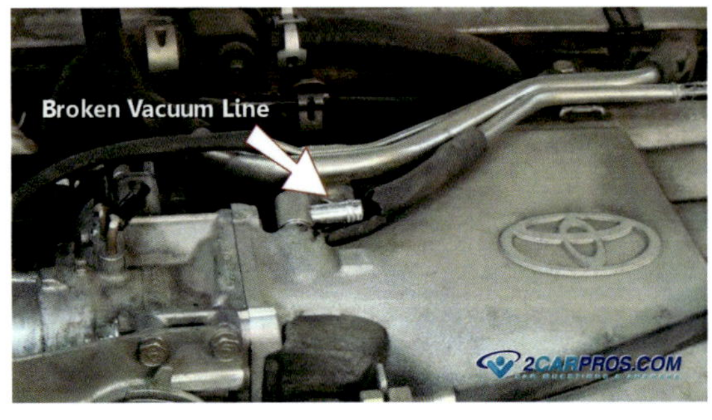

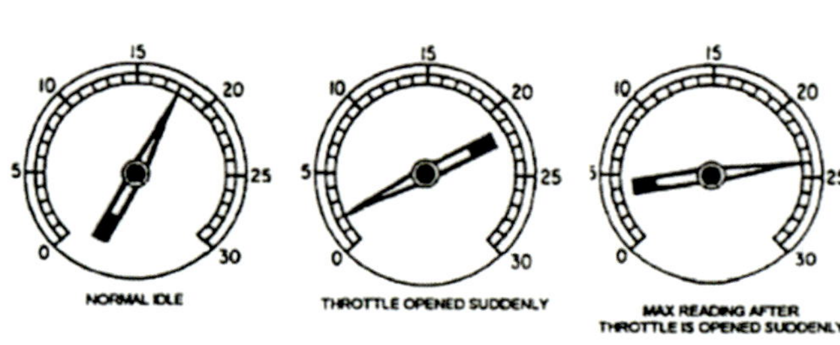

Vacuum Test

- Not all vacuum lines are hooked directly to manifold vacuum. As you gain experience working with engines you will be able to quickly identify vacuum lines and <u>vacuum ports</u> that will allow you to hook up a vacuum gauge for testing purposes

Vacuum Test

You will find that the length and condition of your vacuum hose will affect the accuracy of your vacuum gauge. Old - used vacuum hose can easily collapse and provide a false reading to the vacuum gauge. Longer lengths of vacuum hose will slow down or eliminate quick changes in vacuum

Vacuum Test

A difference of one pound per square inch pressure will register as two inches of mercury. 1psi = 2"hg on the vacuum gauge

Vacuum is created when each piston travels down the cylinder with the intake valve open. It will change drastically as the engine is running at different RPM and load conditions

Vacuum Test / Throttle Plate

A primary factor in manifold vacuum is throttle plate position. When the throttle is closed, manifold vacuum should be higher than when the throttle is open. With an open throttle it is easy for the outside air to get into the engine and there will be less difference in pressure between the outside air and the air found in the intake manifold

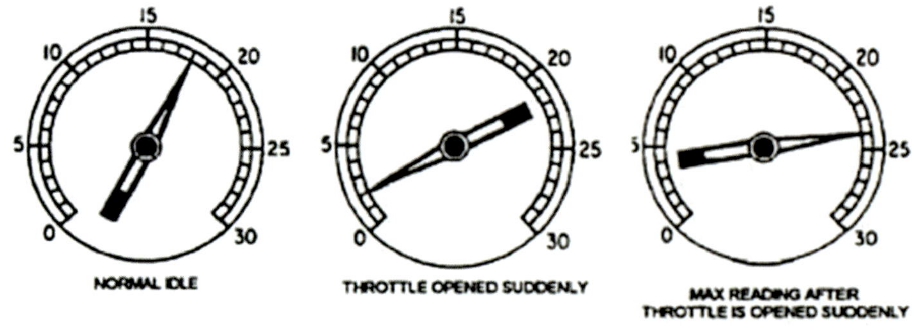

Vacuum Test / RPM

A slow turning engine will produce less manifold vacuum than a fast turning engine. Manifold vacuum will be at its highest when the engine RPM is high, the throttle plate is closed and the RPM are decreasing

Vacuum Vs Compression

If an engine has low compression in all cylinders, it creates a steady, but low, manifold vacuum (Fig 6)

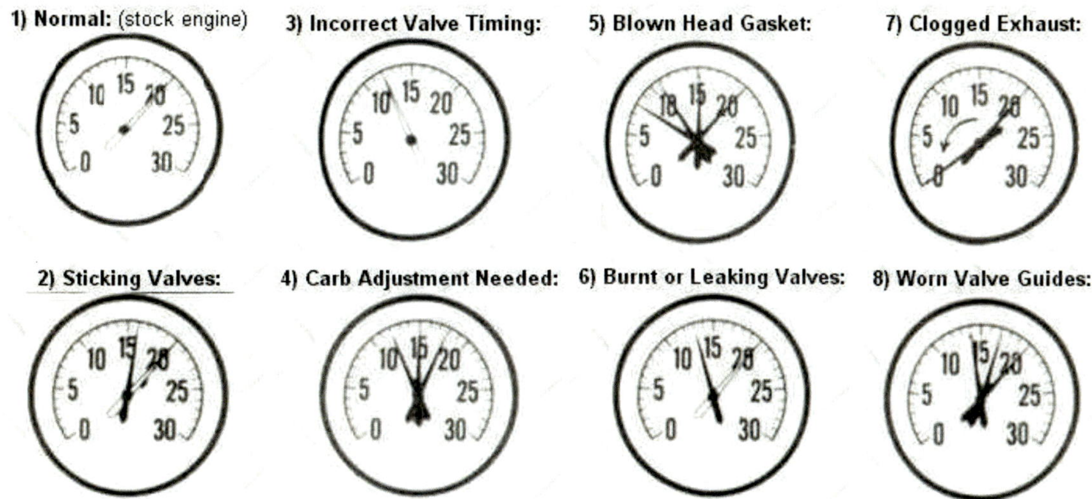

Vacuum Vs Compression

If an engine has low compression in one cylinder, manifold vacuum will be low only during that cylinders intake stroke. This will be observed as a gauge that fluctuates or quickly changes reading (Fig 5)

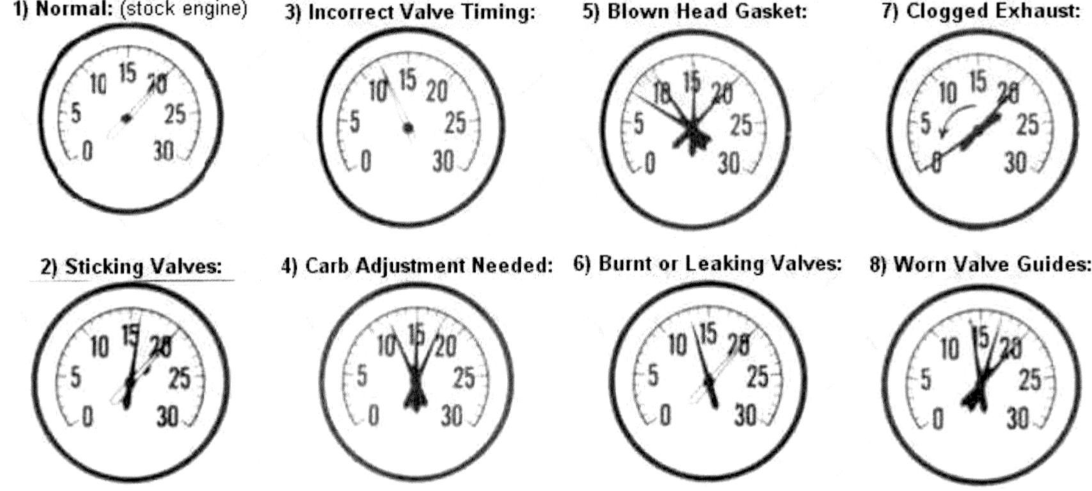

Vacuum Test Tips

- There are a few general guidelines that you can follow when using a vacuum gauge on gasoline engines
    - 1) At idle, the more vacuum the better
    - 2) The highest manifold vacuum will be on deceleration
    - 3) A closed throttle will create more vacuum than an open throttle
- As the engine load increases, the throttle will open wider. This will allow for more air and fuel, create more power, and lower the manifold vacuum reading

Air Control Valve

Finding no cranking vacuum on a gasoline engine does **NOT** mean the engine has poor compression. Many vehicles will use an idle air control valve that opens wide during cranking. The idle air control valve opens an air passage that bypasses the throttle plate. This lets in more air and lowers the cranking vacuum

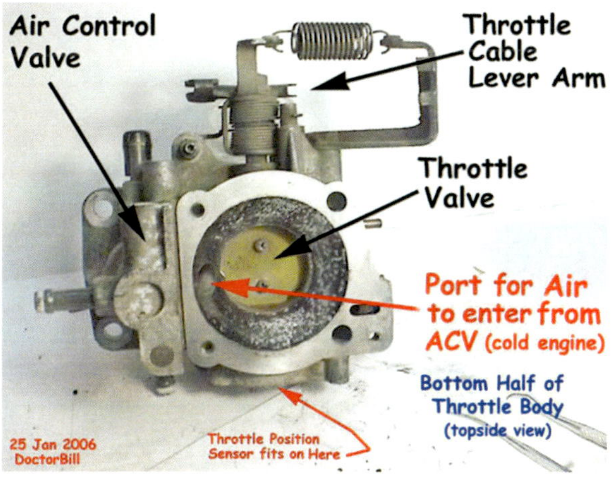

Engine Compression

Low compression in a cylinder can be caused by leaking valves or rings, leaking head gaskets, or other ways that keep the cylinder from sealing during the compression stroke

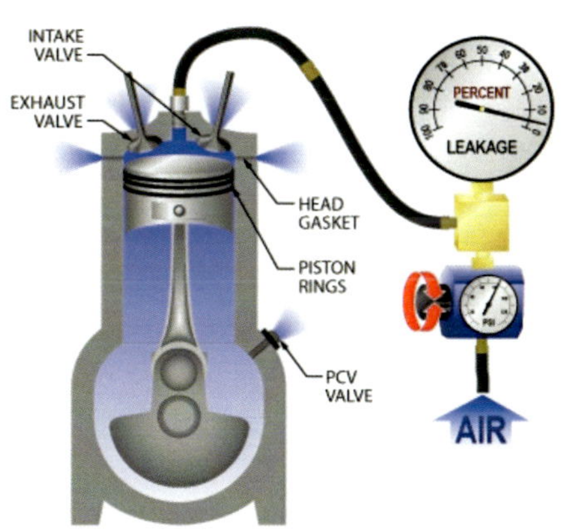

Compression Vs Intake & Exhaust

Engine compression is directly affected by how easily the engine can breathe. If a cylinder can not pull in fresh air it will have low compression. An engine with a restricted exhaust will lose its ability to bring in fresh air

Compression Leaks

A final factor affecting manifold vacuum is leaks in the air intake between the throttle plate, and the intake valves. Leaking vacuum lines, vacuum operated devices, or intake manifold gaskets are common causes for low vacuum

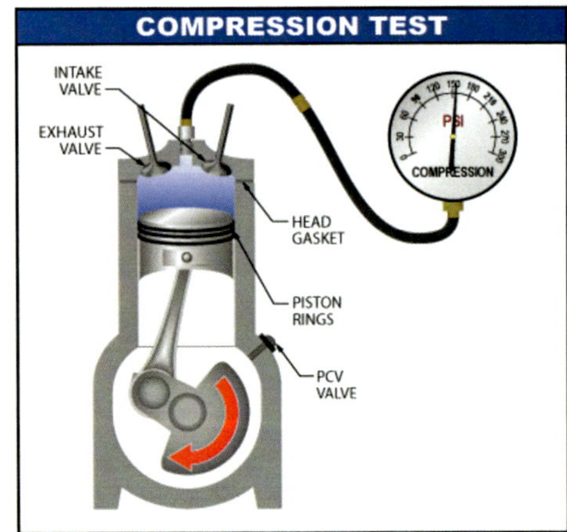

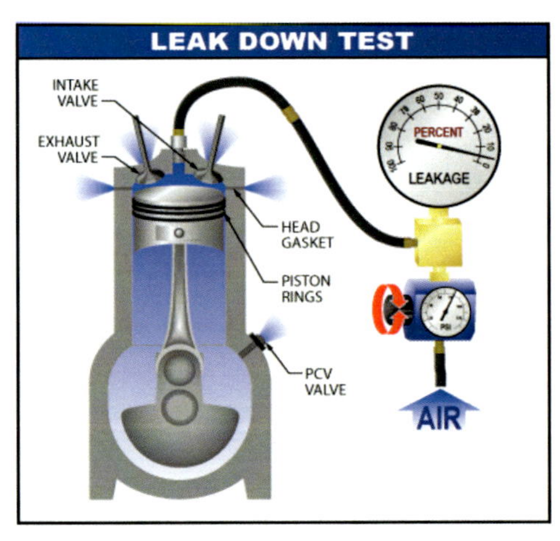

Chapter 3
Ignition Systems
Inboard & Outboard

Inboard Engines Ignition System

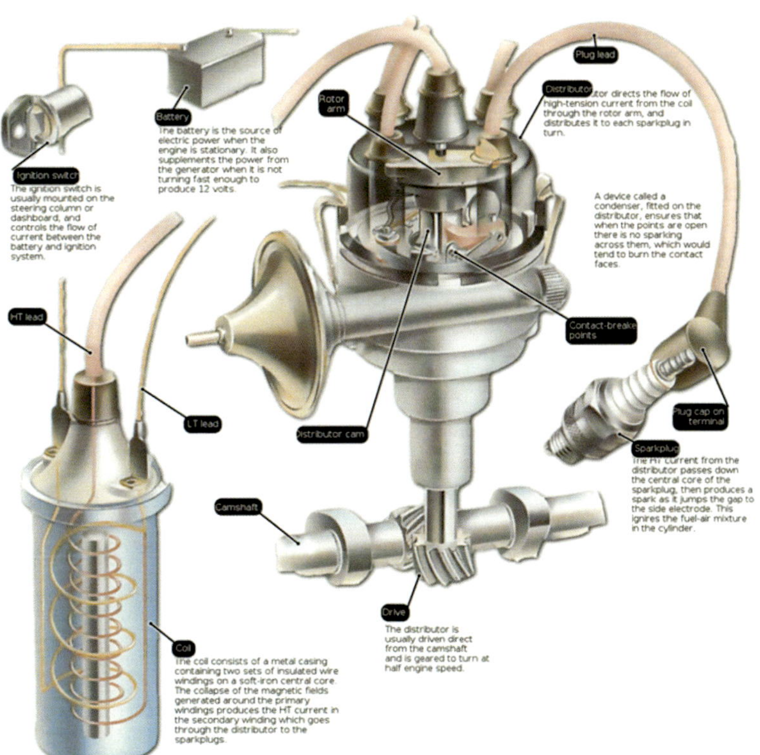

Ignition System

- There are three types of ignition system for gasoline engines

  - **Electro-Mechanical system** with: Points, Condenser, Rotor and Ignition Coil
  - **Magneto-Ignition system** with an electromagnet instead of points and condenser to produce the input signal to the coil
  - **Electronic Ignition system** with a microprocessor integrated into the ECM

Electro-Mechanical Ignition

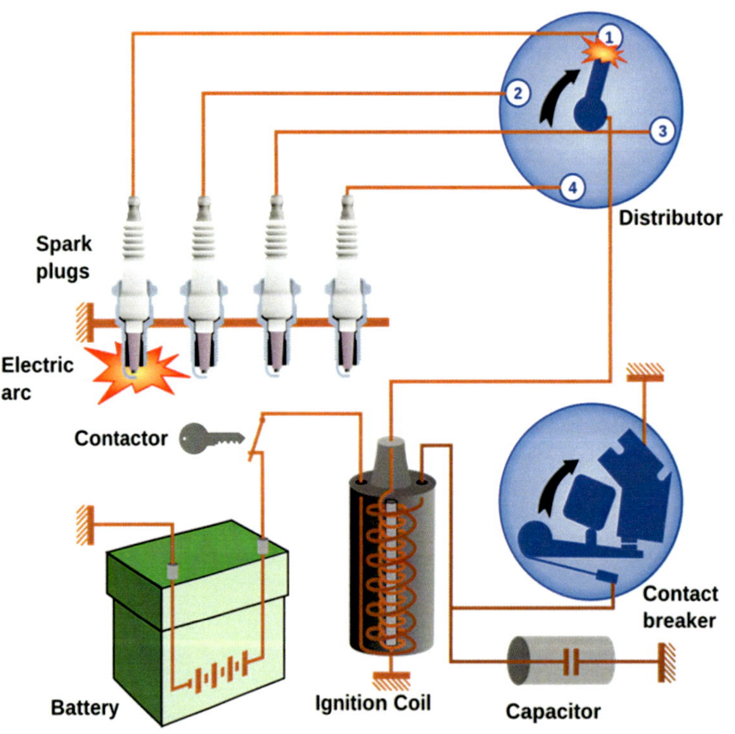

Electro-Mechanical Ignition

This system has two circuits. The primary circuit (12 V) controls the input current to the coil. (Battery, Ignition Switch, Points and Condenser)

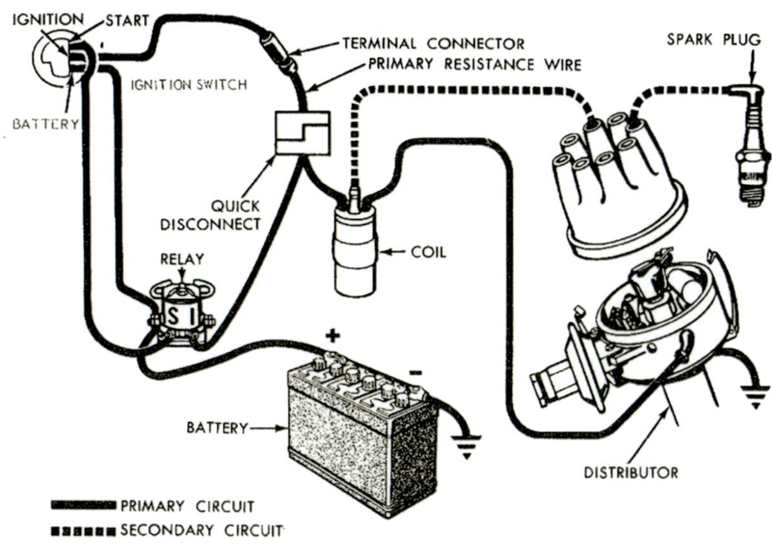

Electro-Mechanical Ignition

The secondary circuit (High Tension 12000 V), starts at the output of the coil , enter in the distributor and then by means of the rotor, it is sent to the spark plugs

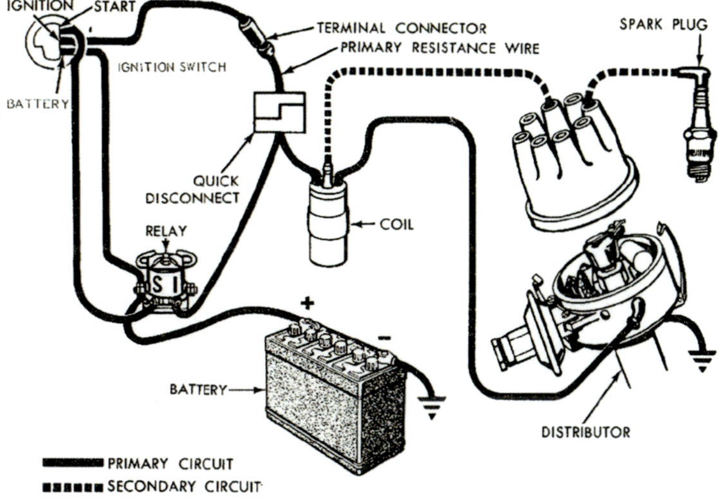

How the ignition System Works?

The sequence of the ignition firing begins with the points (or the contact breaker) closed.

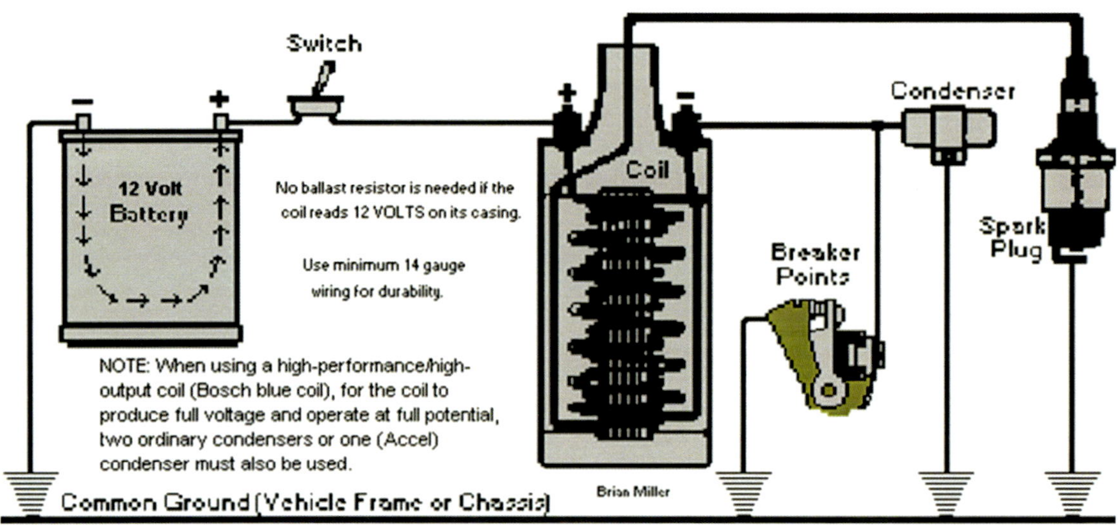

How the ignition System Works?

A current will flow from the battery (+), through the ignition switch, then through the coil primary and across the closed breaker points (-)

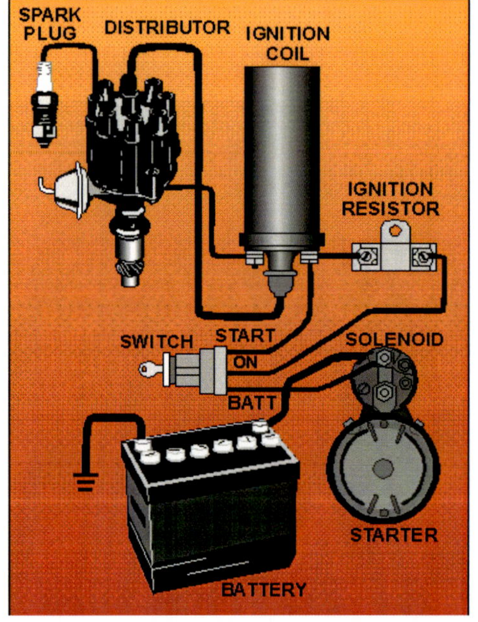

How the ignition System Works?

This current produces a magnetic field within the coil's core. This magnetic field will form the energy reservoir that will be used to drive the ignition spark

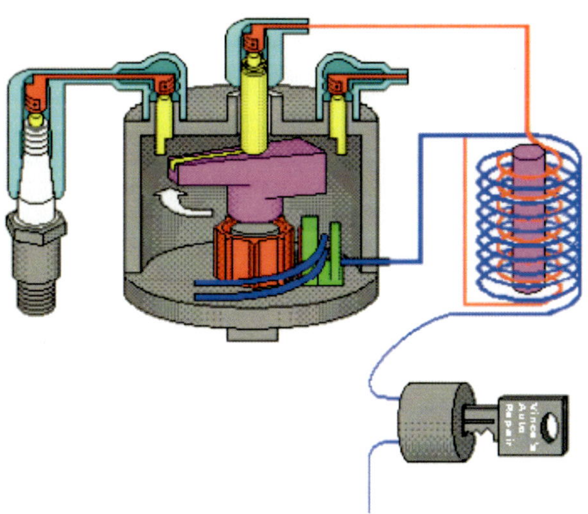

How the ignition System Works?

The ignition coil's secondary windings are connected to the distributor cap. A turning rotor that is located on top of the breaker cam within the distributor cap, connects the coil's secondary windings to one of several wires that lead to each spark plug

How the spark is interrupted?

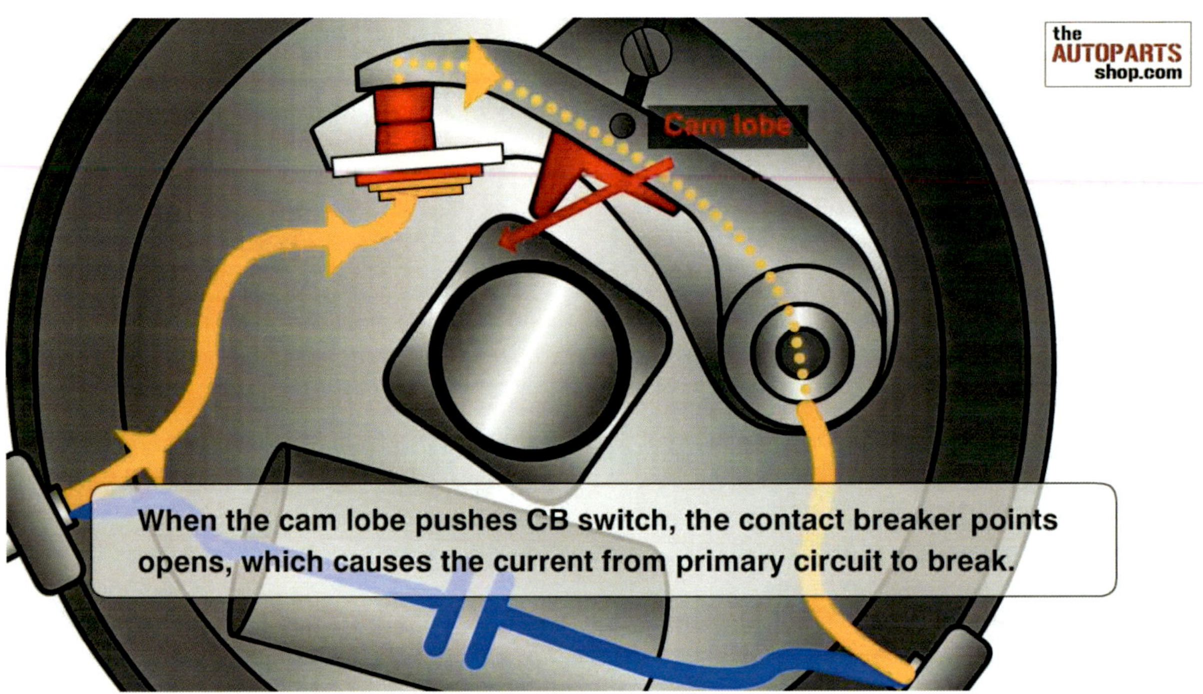

Mechanically Timed Ignition

- The heart of this system is the distributor which contains:
  - (E) A rotating cam which runs off the engine's drive
  - (B) A set of breaker points
  - (F) A condenser
  - A rotor
  - A distributor cap.

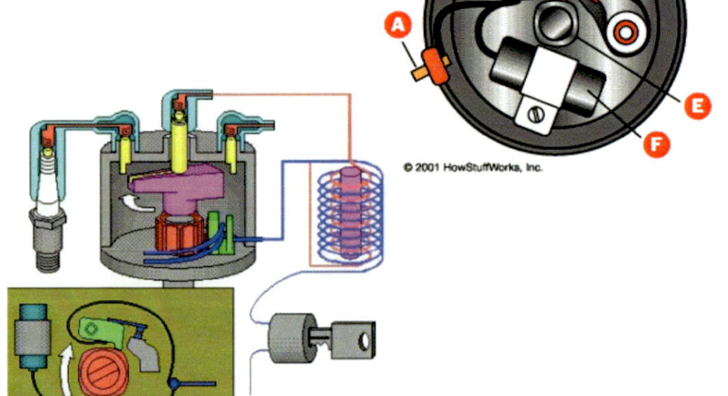

Mechanically Timed Ignition

- External to the distributor is the:
  - Ignition coil
  - Spark plugs and then the wires that link the spark plugs and the ignition coil to the distributor.

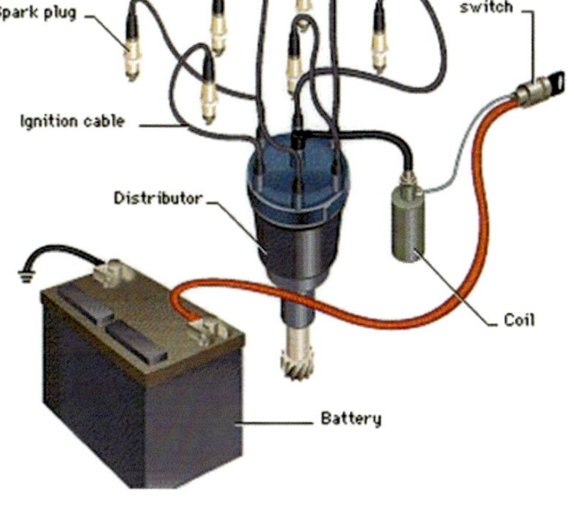

The Ignition Coil

An ignition coil is really two separate coils of wire. They are known as the Primary (low voltage) coil and the Secondary (high voltage) coil

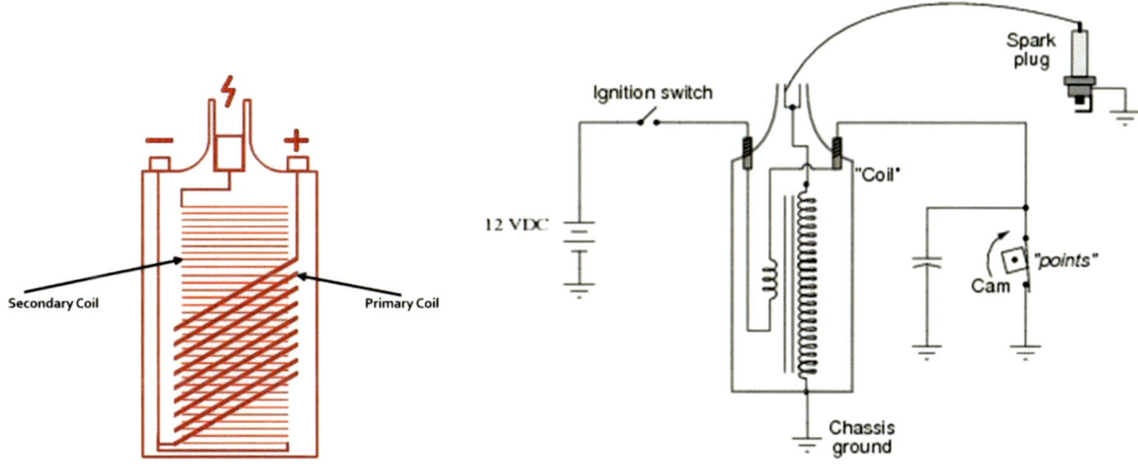

The Ignition Coil

As the magnetic field in the primary windings collapses it moves rapidly past the secondary coil windings. This will induce a very high voltage into the secondary coil windings.

The Ignition coil

For an ignition coil, one end of the windings of both the primary and secondary are connected together. This point is connected to the battery (+). The other end of the primary (-) is connected to the points within the distributor

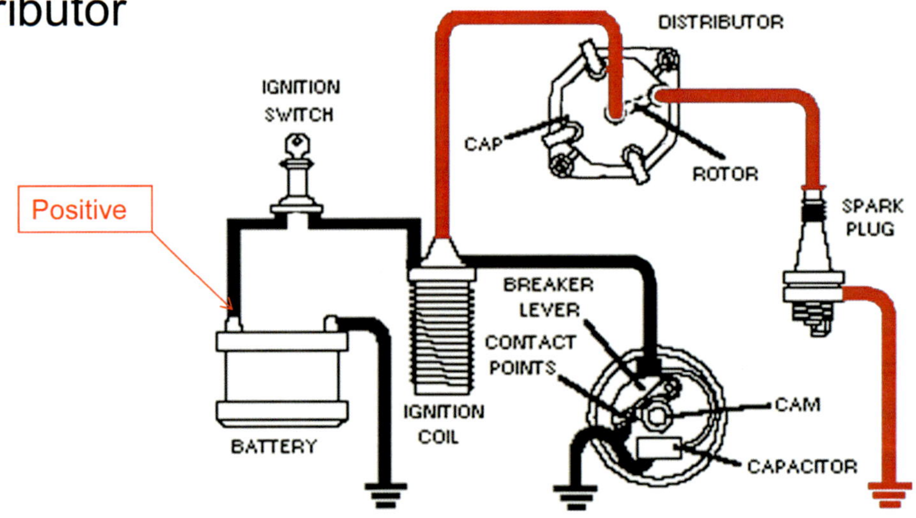

The Ignition coil

Then the other end of the secondary is connected, via the distributor cap and rotor, to the spark plugs

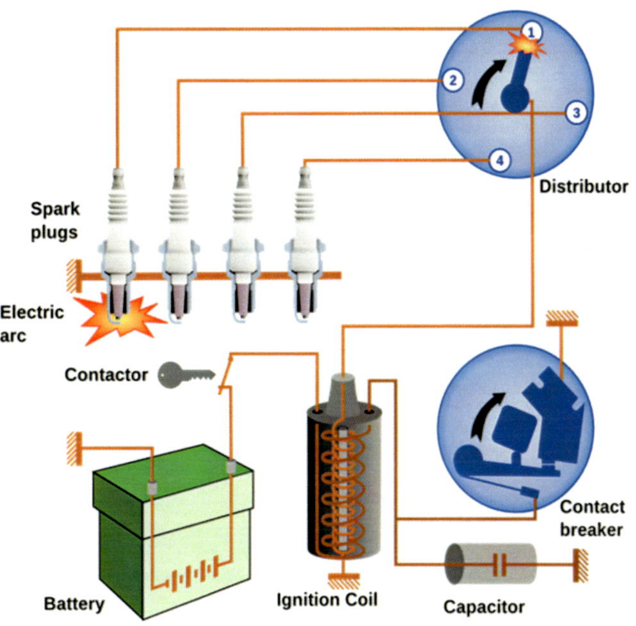

Failure to Start

Here are the basic reasons of how the air/fuel can fail to start burning and cause the gasoline engine to not start.

- The compression can be too low inside the cylinder
- There can be not enough fuel in the cylinder
- There can be too much fuel in the cylinder
- The primary coil does not turn ON
- The primary coil does not turn OFF
- There is a short in the secondary ignition circuit
- There is an open in the secondary ignition circuit

Magneto Ignition System

- There are two types of Magneto Systems Breaker Point Ignition

- Solid State (Electronic) Ignition

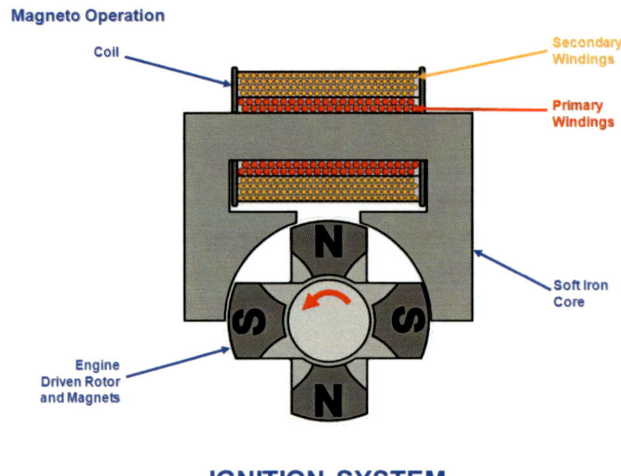

Magneto Ignition System

An **ignition magneto**, or **high tension magneto**, is a magneto that provides current for the ignition system of a spark-ignition engine. It produces pulses of high voltage for the spark Plugs

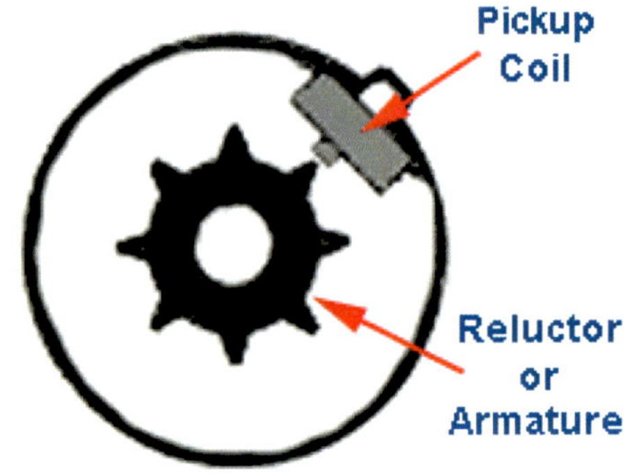

Electro-mechanical Vs Magneto

- With a simple kit, the electro-mechanical system can be converted into an electronic system or Magneto system . Just the internal parts of the distributor should be replaced

Electronic Distributor Advantages

- Less moving part hence minimal wear and tear hence better performance
- In point distributor, sparking at the points burns the contacts and wastes energy, Electronic ignition systems allow most of the available energy to reach the spark plug and improve combustion
- Better starting and smoother running

Electronic Distributor Advantages

- No need to check or adjust and clean contact breakers
- No capacitor to fail or go weak
- More water resistant

Advantages of Magneto Ignition Systems

- No battery required - the permanent magnet moving at speed past the coil induces the necessary high tension current for the spark
- The quality of the spark improves with engine speed. (because there is a higher speed of cut of the magnetic lines of force across the conductors in the coil)
- Less wiring compared to other ignition systems
- Simple layout and operation

Magneto Ignition Systems

Every time the permanent magnet passes the coil, it induces an electrical current in the coil. Because the coil has so many turns, the voltage of the induced current will be high - high enough to produce a spark at the spark plug

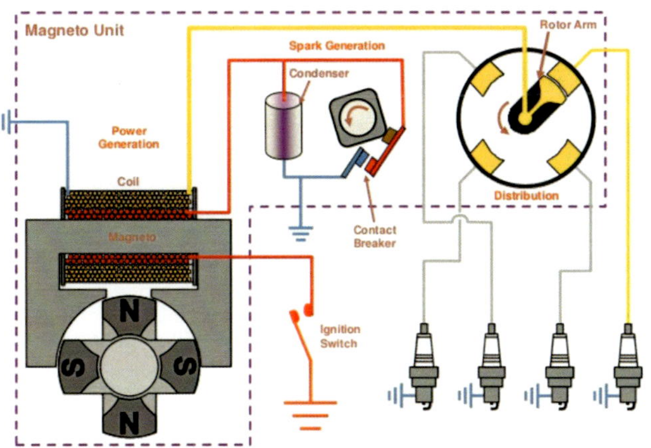

IGNITION SYSTEM – Magneto System

Electronic Ignition System

The function of the points and condenser (Interrupt the ground signal to the primary circuit) is made by the ECM

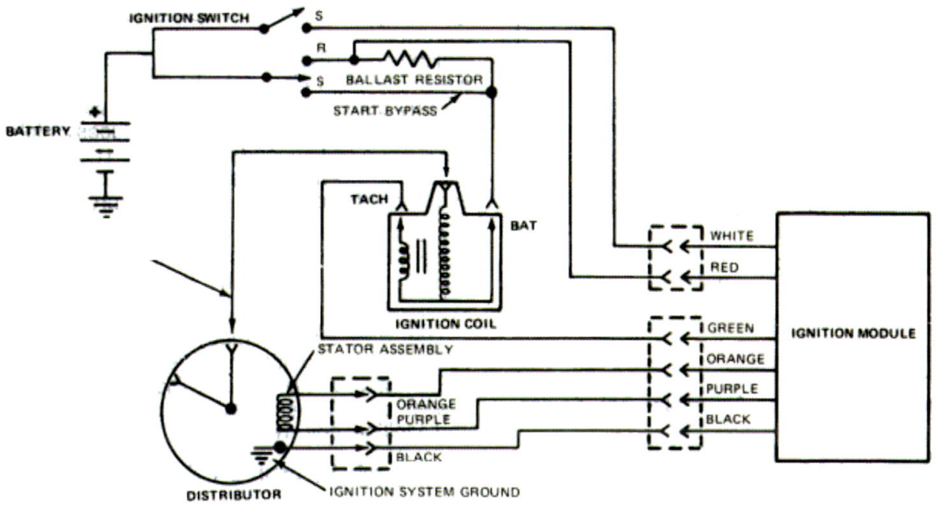

Electronic Ignition System

The ECM just interrupts the ground wire of the primary circuit , doing the same function of the points and condenser in the mechanical system

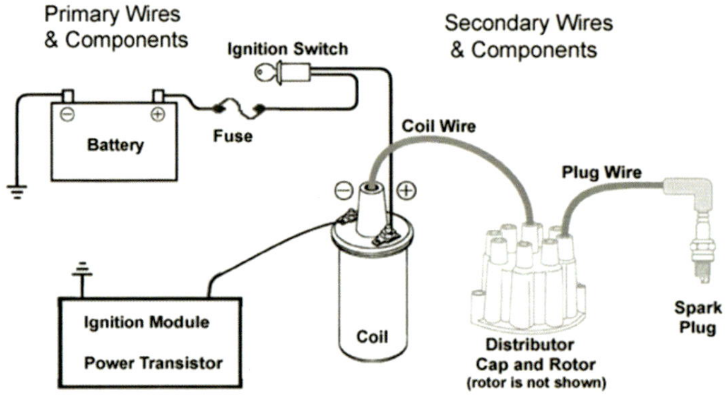

Primary wires are low voltage. They have thin insulation and are skinny.
Secondary wires have more insulation to resist the high voltage.
Secondary wires are much thicker than the primary wires.

Timing Inboard Engines

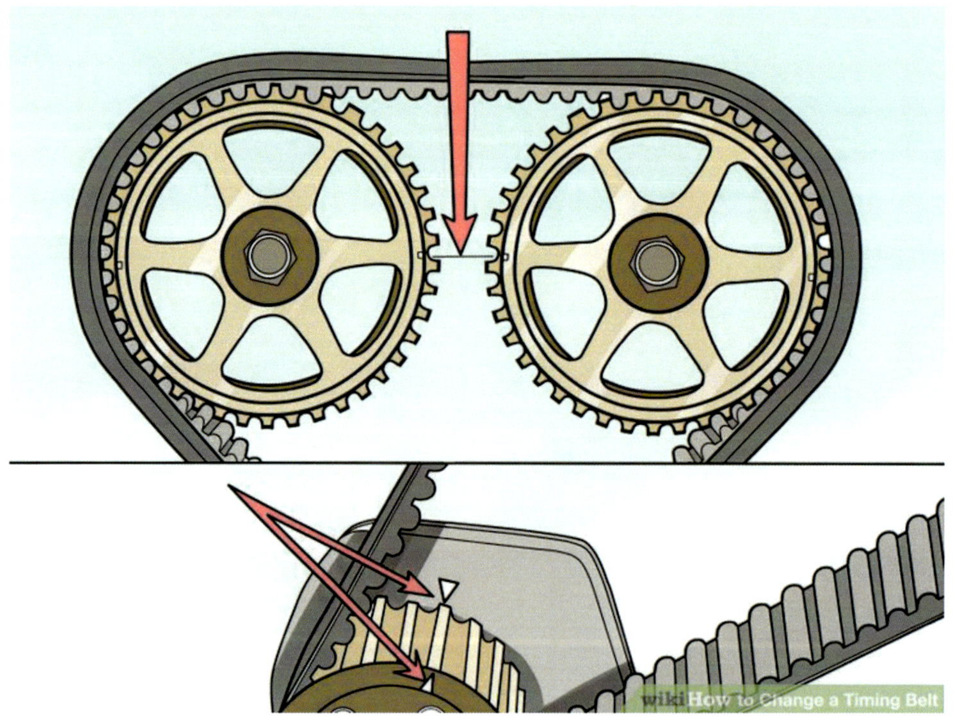

Cylinder Numbering

- In a straight engine the cylinders are numbered, starting with #1, usually from the front of the engine to the rear

- In a V engine, cylinder numbering varies among manufacturers. Generally speaking, the most forward cylinder is numbered 1

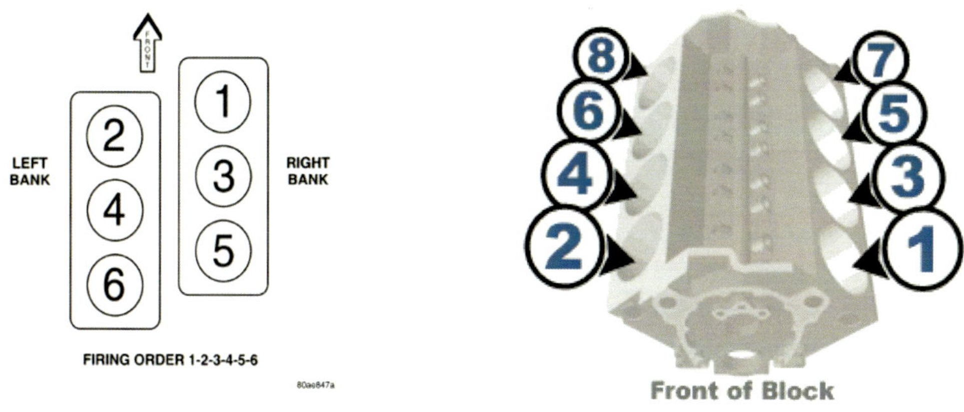

Block Firing Order

The **firing order** is the sequence of power delivery of each cylinder in a multi-cylinder reciprocating engine

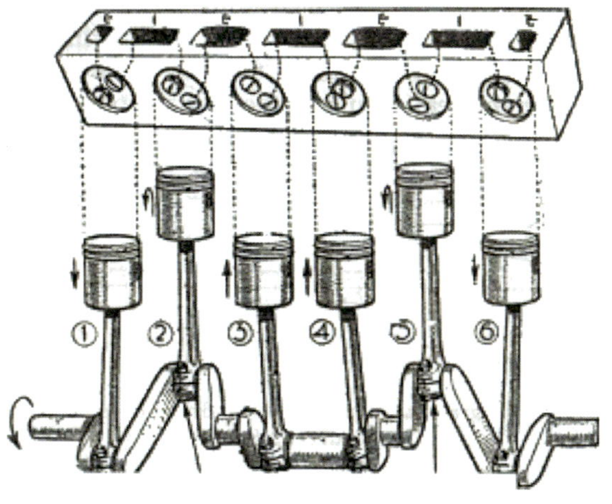

Firing Order

The order in which combustion occurs in the cylinders of an engine. Also the order in which spark is distributed to the plugs by the distributor

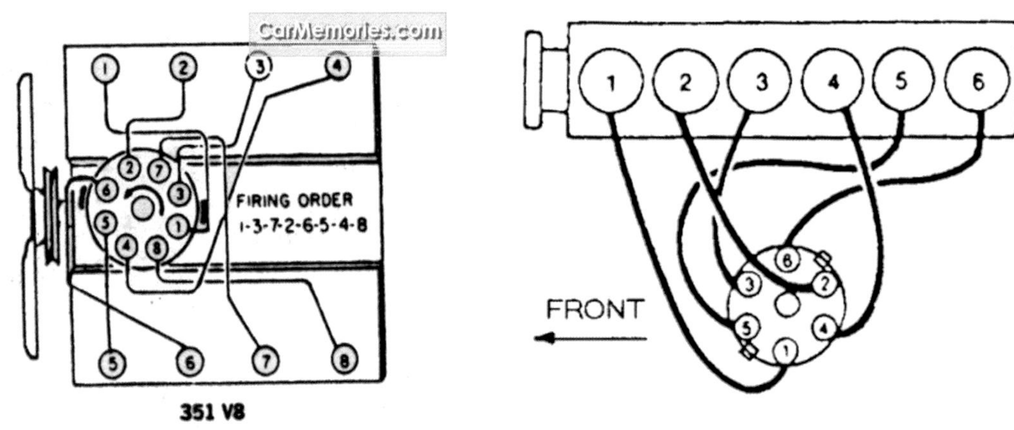

Firing Order (Gasoline)

In a gasoline engine, the correct firing order is obtained by the correct placement of the spark plug wires on the distributor

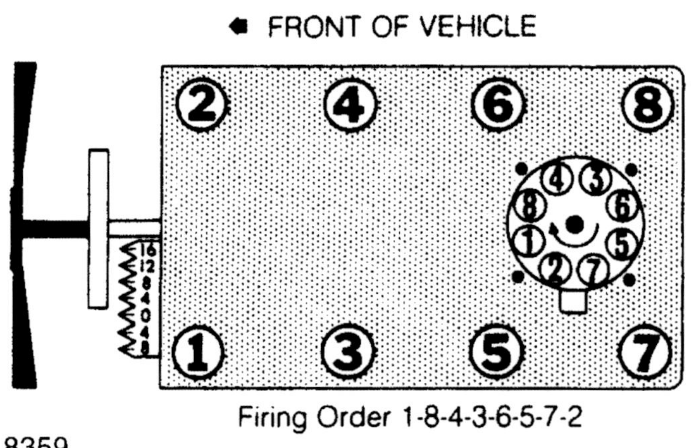

Firing Order 1-8-4-3-6-5-7-2

Firing Order (Gas)

In a gasoline engine, the correct firing order is obtained by the correct placement of the spark plug wires on the distributor

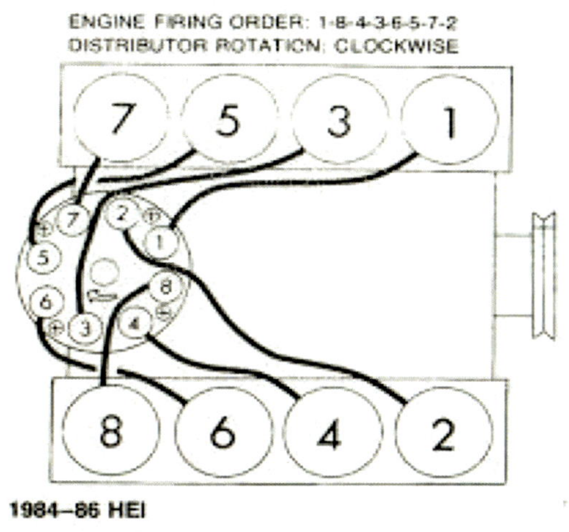

Clockwise Rotation (Gas)

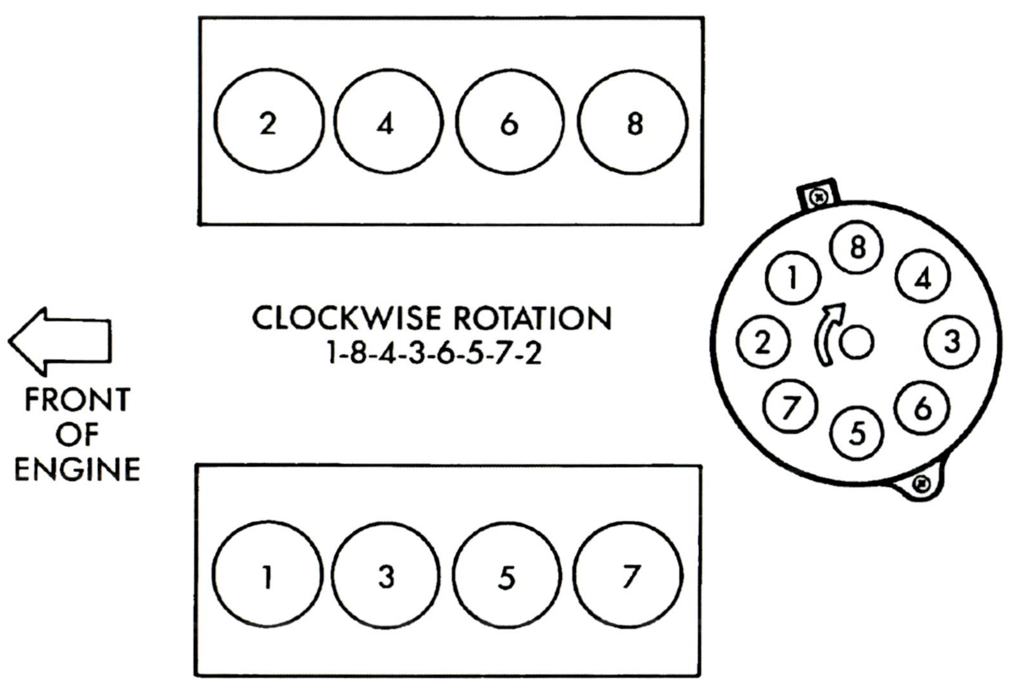

Counterclockwise Rotation

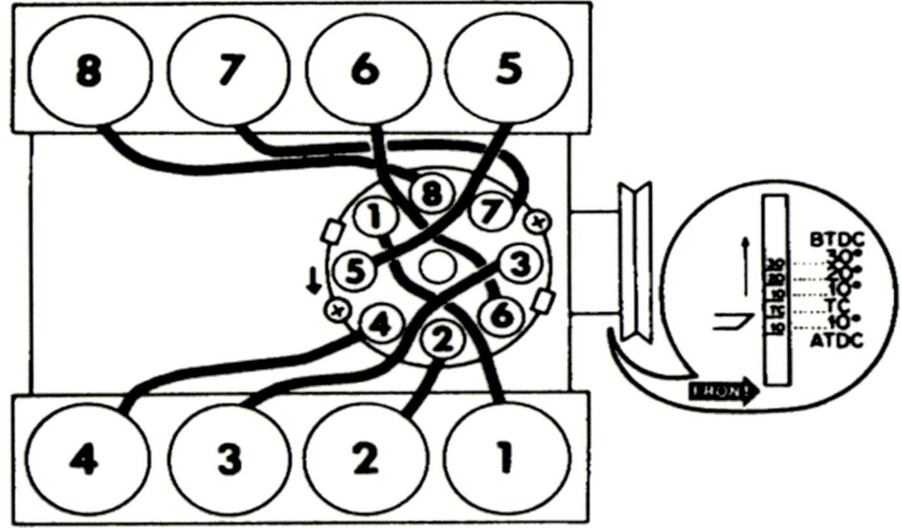

**1975–88 8-460; 1979–88 8-302
firing order: 1-5-4-2-6-3-6-7-8
distributor rotation: counterclockwise
NOTE: Squares are the latches on 1975–76 models;
circles are the latches on 1977–88 models**

Timing Marks

A **timing mark** is an indicator used for setting the timing of the ignition system of an engine, typically found on the crankshaft pulley (as pictured) or the flywheel, being the largest radius rotating at crankshaft speed and therefore the place where marks at one degree intervals will be farthest apart

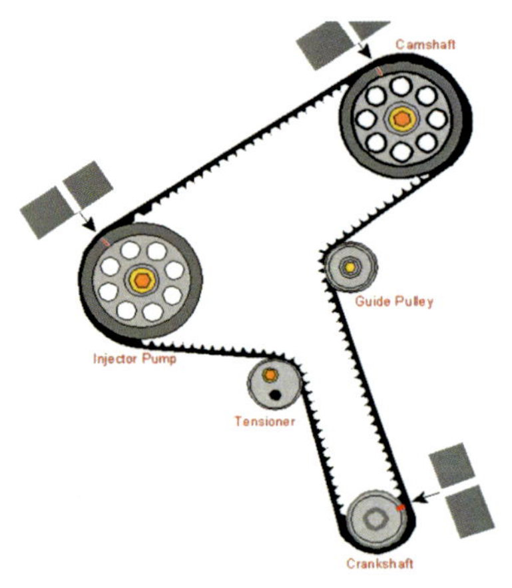

Timing mark

A **timing mark** is an indicator used for setting the timing of the ignition system of an engine, typically found on the crankshaft pulley (as pictured) or the flywheel, being the largest radius rotating at crankshaft speed and therefore the place where marks at one degree intervals will be furthest apart

Timing Mark Adjustment

On older engines it is common to set the timing using a liming light, which flashes in time with the ignition system (and hence engine rotation), so when shone on the timing marks makes them appear stationary due to the stroboscopic effect.

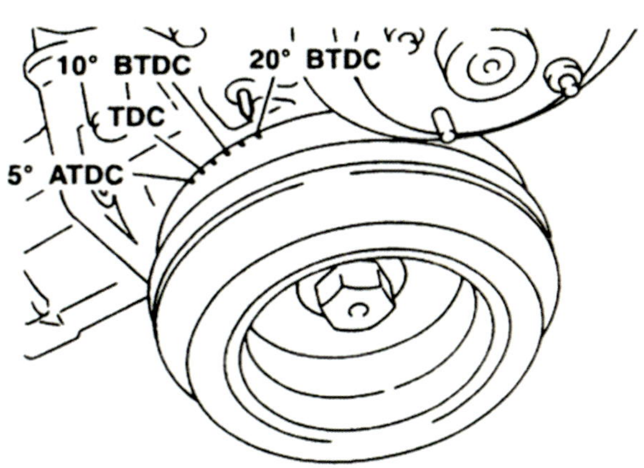

Timing Mark Adjustment

The ignition timing can then be adjusted to fire at the correct point in the engine's rotation. The timing can be adjusted by loosening and slightly rotating the distributor in its seat

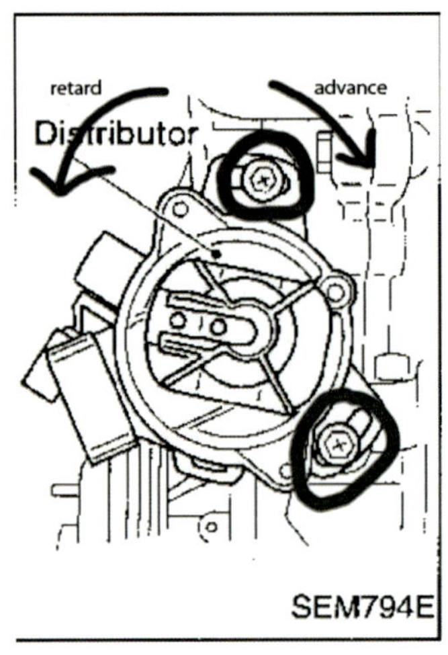

Timing Mark Adjustment

Modern engines usually use both Cam and Crank sensor directly connected to the engine management system

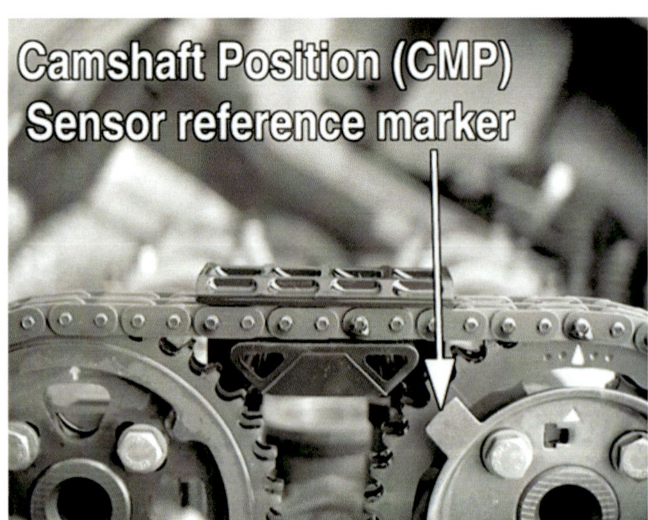

Timing light

The apparent position of the marks, frozen by the stroboscopic effect, indicates the current timing of the spark in relation to piston position. On most automotive engines, the timing is set based on the #1 cylinder

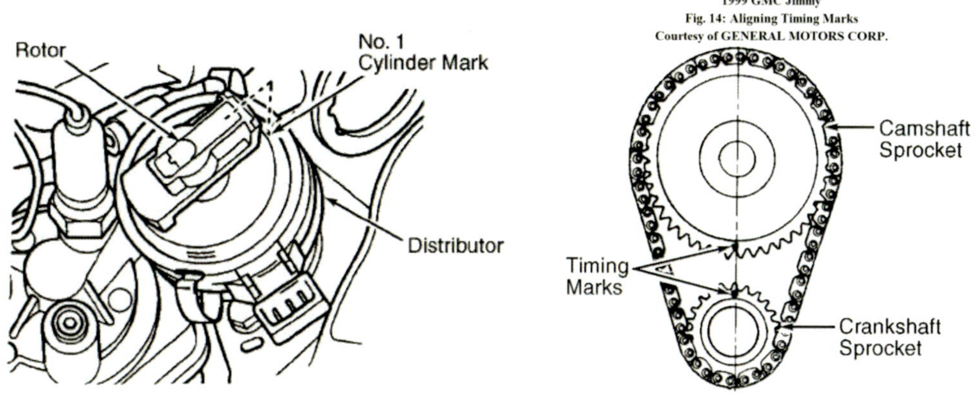

Timing belt

Timing belt is a toothed belt that connects the engine crankshaft to the camshaft or camshafts

Timing belt

It has teeth to turn the camshaft(s) synchronized with the crankshaft, and is specifically designed for a particular engine. In some engine designs, the timing belt may also be used to drive other engine components such as the water pump and oil pump.

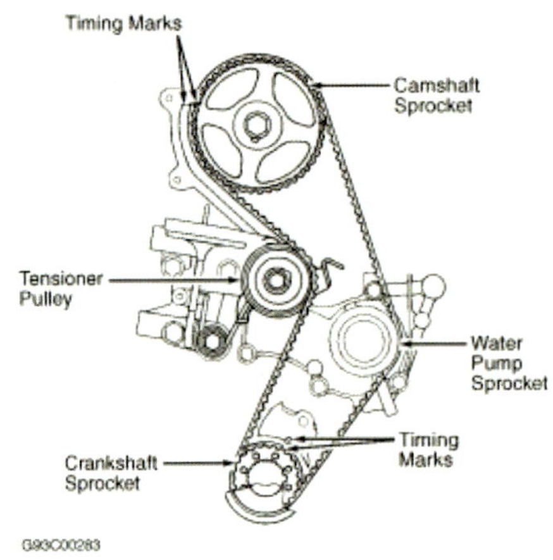

Timing Belt Replacement Intervals

Your timing belt should be replaced every 50-70,000 miles

Timing Belt Replacement Intervals

In an engine, the valves and piston share the same air space. They never touch, unless your timing belt breaks or skips, and this is a catastrophic failure that requires removing the head and replacing bent valves

Timing Chain

The timing chain keeps internal engine components in sync with each other by causing valves to open and close at the proper time

Loose Timing Chain

If the timing chain_is loose or improperly adjusted, it may "jump time" (skipping a tooth or more, usually on the cam gear). This results in a loss of synchronization and engine performance

Timing Belts Vs Timing Chain

- Belts are quieter
- Are less expensive
- Are mechanically more efficient
- Considerably lighter
- Belts do not require lubrication
- a broken timing belt can cause expensive damages

- Chains are More durable
- A timing chain is noisier
- Less efficient, and more expensive

- A sloppy timing chain can result in poor running, valve clatter, and loss of power

Outboard Engine Ignition System

Two Stroke Ignition System

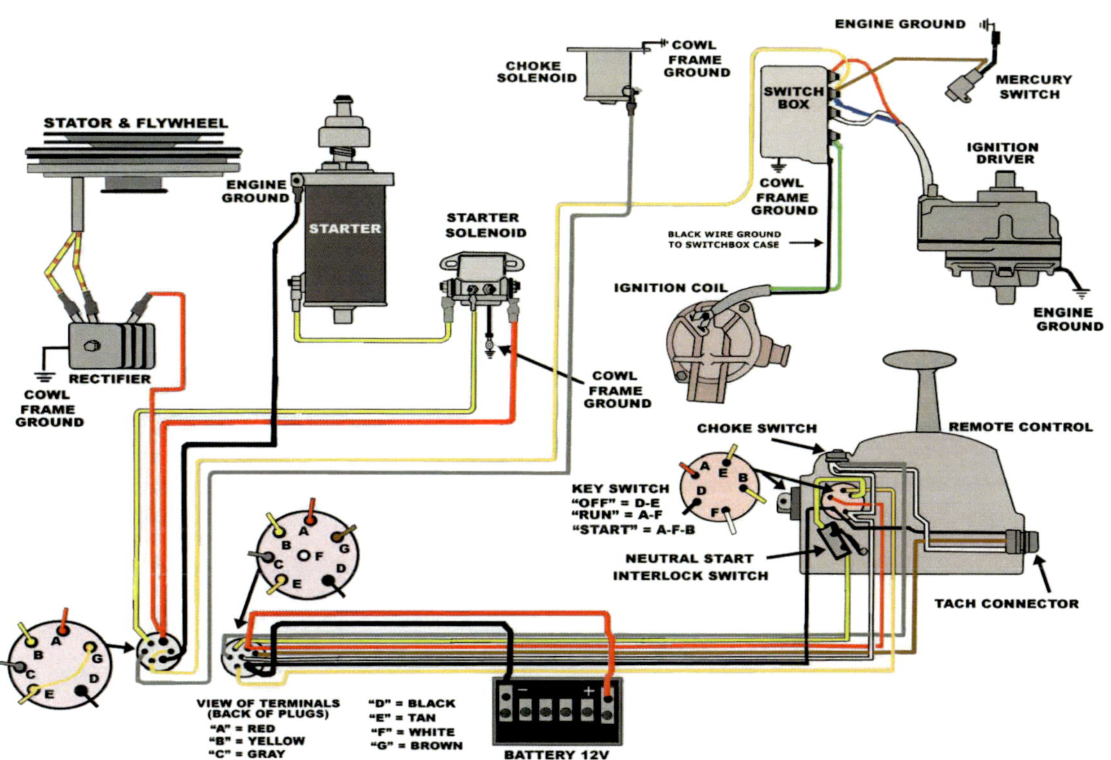

Types of Systems

- Following two types ignition systems used most often in outboards
    - Conventional flywheel magneto system with breaker point contacts operated by a crankshaft lobe
    - The flywheel magneto system with electronic ignition module (capacitor discharge type)
- Actually the majority of the outboards use an electronic system controlled by an ECM

Electromagnetism & Generators

- If we wound many loops of wire into a coil, turned coil in magnetic field, a much larger voltage and current would be produced. The amount of voltage and current produced by generator based on three things:
    - 1)The number of turns in coil and diameter of wire
    - 2)The strength of magnetic field
    - 3)The speed at which wire coil passes by magnets

Flywheel, Stator & Trigger

The flywheel works in sync with the stator and trigger providing both the electrical fields to charge energy into the system and a good indication of crankshaft position

Stator Operation

- The stator on outboard motors is located underneath the flywheel. The stator plays two roles.

  - The flywheel contains magnets mounted in a 360 degree configuration. As the flywheel rotates, these magnets pass over the stator coils, which in turn, generates an electrical current. One set of the coils on the stator feeds the ignition system. Another set of coils provide electrical current to charge the starting battery.

Stator Operation

The stator sits stationary and does not move. The stator has a series of coils with most of them used to generate electricity for the charging system.

Stator Operation

The stator has separate coils designated as charge coils for the ignition system which function independently. The charge coils are what power the entire ignition allowing it to work properly

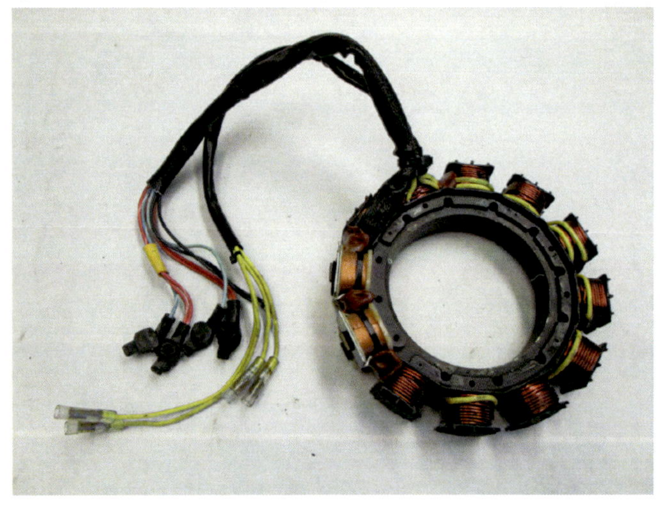

The Voltage Regulator

- The regulator is basically what the name says. It regulates voltage output from the stator to the starting battery. Without it, the stator would output too much voltage to the battery resulting in overcharging and damage to the battery

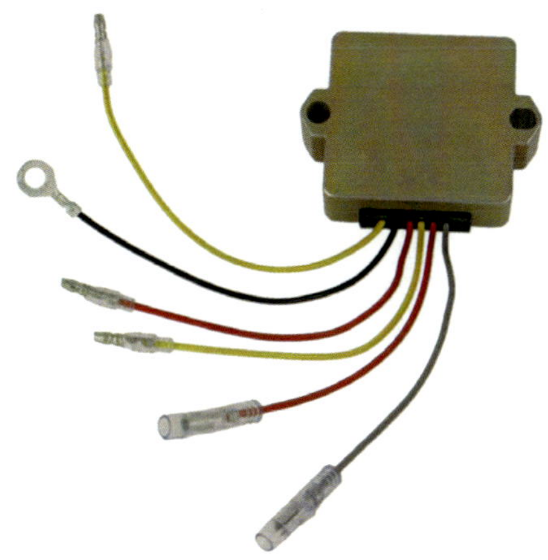

The Voltage Regulator

- The regulator is basically what the name says. It regulates voltage output from the stator to the starting battery. Without it, the stator would output too much voltage to the battery resulting in overcharging and damage to the battery

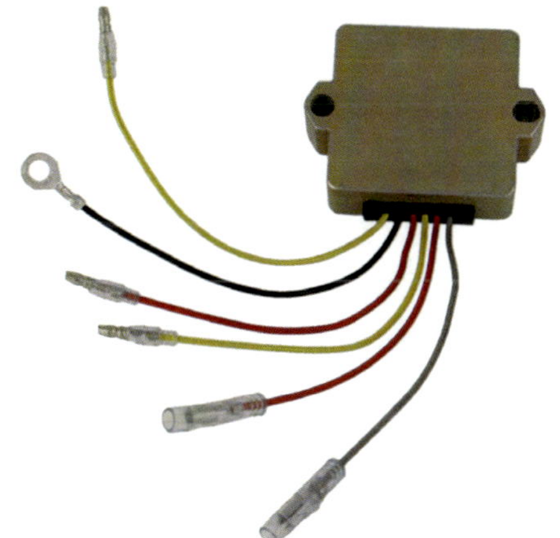

Voltage Regulator / Rectifier

- Most regulators also contain a rectifier circuit which is used to sense electrical pulse and provide a signal to the tachometer to read the motor's rpm. Additionally ,the rectifier supply DC power to keep the battery fully charged

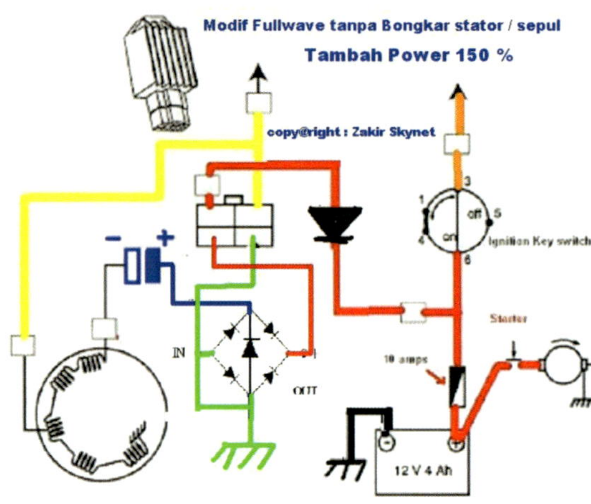

The Trigger / Timerbase

- The trigger or timerbase (term varies by manufacturer) is similar to a distributor on an automotive application. This component is also mounted under the flywheel and also uses the flywheel for it's operation

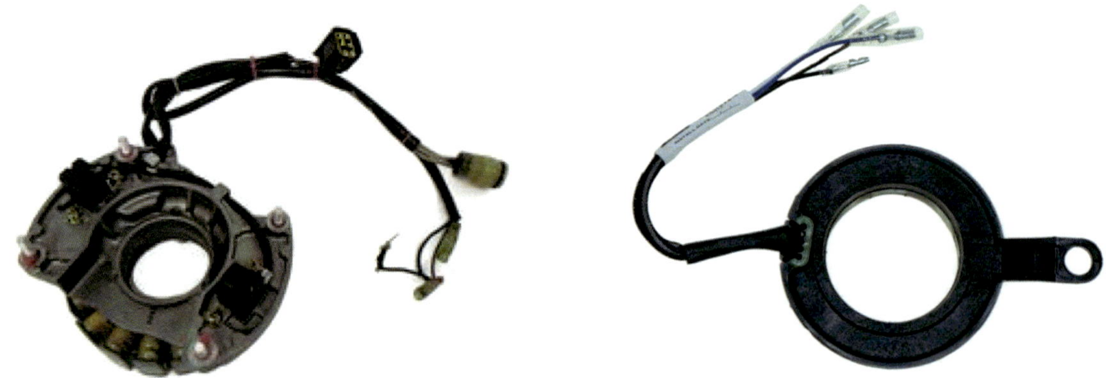

The Trigger / Timerbase

The center hub on the flywheel also contains magnets mounted in a 360 degree configuration. Instead of generating electrical current though, these magnets simply tell the trigger/timerbase what position the crankshaft is at. As the magnets pass by the pickup sensors, the trigger/timerbase sends a signal to the power pack, telling it the precise moment to release its stored energy to the specific cylinder

The Trigger / Timerbase

- The spark timing is controlled exclusively by the trigger/timerbase. As the component rotates with increase or decrease of the throttle, so does the spark timing. Smooth and accurate movement of this device is critical to a good running motor

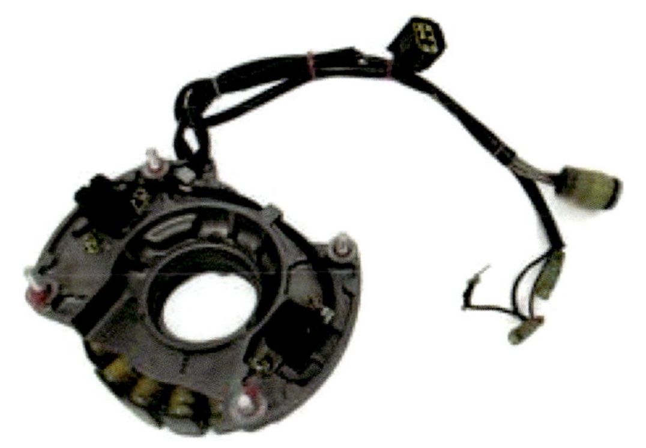

The Trigger Coil

The trigger assembly senses crankshaft position in a way communicating when the system should fire. This would be similar to the function of a pick-up coil in a distributor or a crank angle sensor in some of the newer cars

The Power Pack / Switch Box

The function of a power pack is two fold. First, it stores energy created from the stator generating electrical current directly to it (ignition coils). Secondly, it releases that stored energy per instructions from the trigger/timerbase. Hence the term "capacitive discharge ignition". Where does that released energy go? To the ignition coils of course

The Ignition Coil

This is the most simplistic component in the ignition system. The coil's function is to simply boost the energy output from the power pack. Basically it is a transformer

The Ignition Coil

For example, the output from the power pack might be something like 300 volts or so. The ignition coil boosts that current to very high voltage (sometimes as high as 10,000 to 15,000 volts depending on the application) which is required to fire the spark plugs and burn the fuel in the cylinder

The Ignition Coil

The coil output can have a direct impact on how well the fuel mixture burns. There's not really any worries for high output (actually the higher the better), but with low output, the fuel mixture burns poorly, carbon buildup increases drastically, and performance obviously suffers

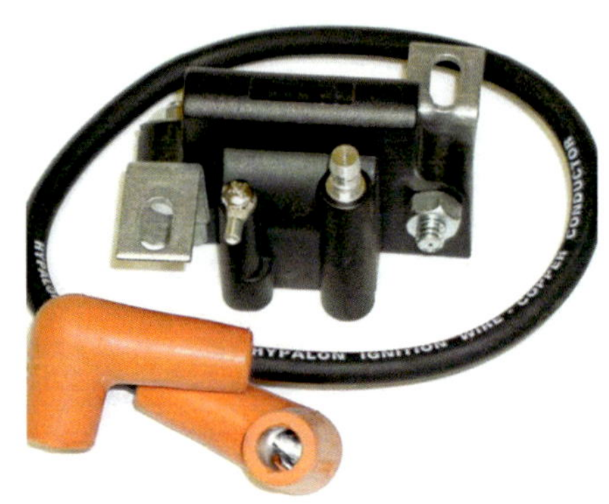

Spark Advance

It is important to note that due to mechanical movements at higher RPM's, the timing of when the spark plugs fire will need to change to compensate for the fuel burning at the correct time

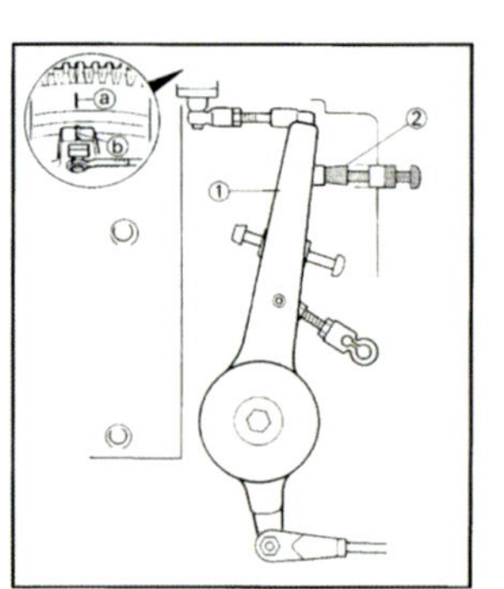

Timing Adjustment

This is why the timing needs to be advanced or retarded to the correct setting. The trigger does move for this reason but its movement is limited by the throttle mechanism which has an adjustment

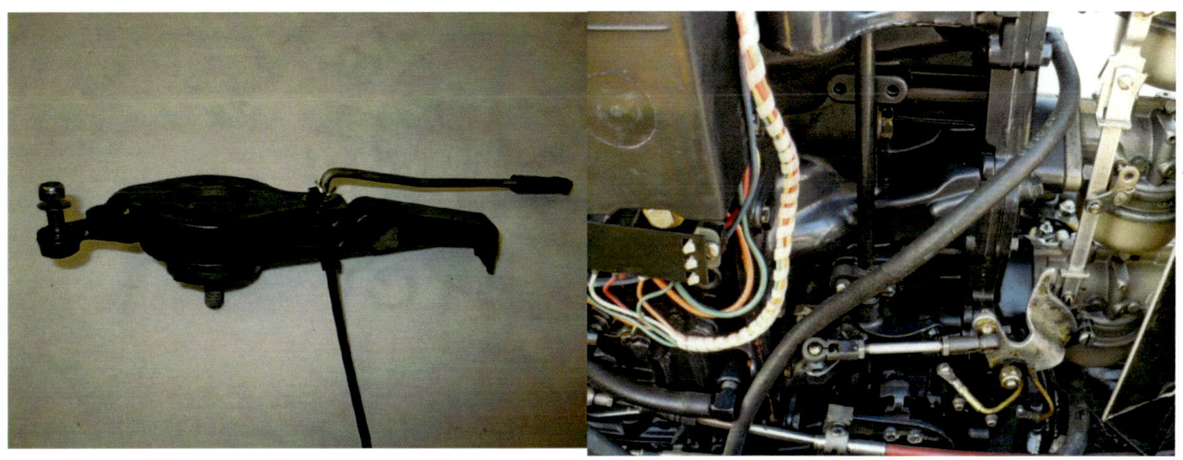

Timing Adjustment

The adjustments on the trigger are important to a good running outboard motor either at idle speed or wide open throttle

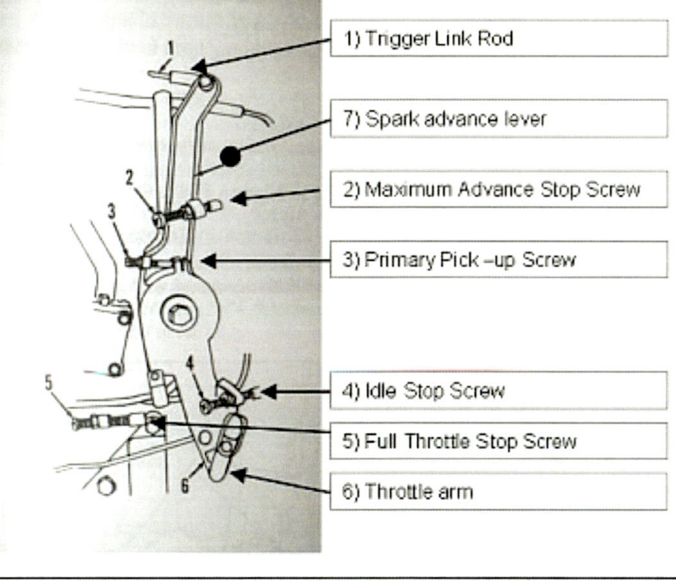

ADJUSTING OUTBOARD IGNITION TIMING

- Disconnect the spark plug wire from the spark plug over Cylinder 1 and connect the timing light to the wire

- Run the engine at full throttle.

ADJUSTING OUTBOARD IGNITION TIMING

- The timing mark should align with the recommended degree mark on the flywheel

- According with the manufacturer recommendations

ADJUSTING OUTBOARD IGNITION TIMING

A reading of plus 1 degree or minus 1 degree generally acceptable. If timing off by more than 1 degree, should be adjusted. To adjust timing, stop the engine and turn timing adjustment screws to either advance or retard timing. Then, restart engine and recheck the timing with timing light

Spark Plugs

- First check to see that the plugs look the same on the electrode.

- If one plug is completely different than the other, check for air leaks or one carburetor not working correctly.

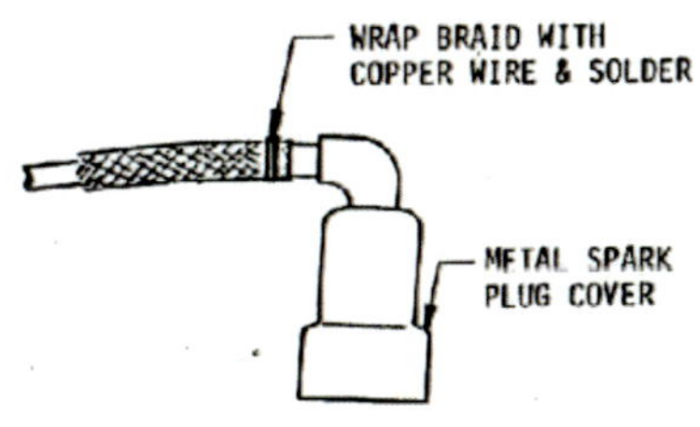

Spark Plug/ Ignition Coil

- Also check for an ignition problem if one spark plug is black and fouled showing that it was not firing
- If the plugs look even or if it is a single cylinder engine, check the coloring on the electrode

Spark Plug Condition

- If it's black the mixture is too rich or your plugs are fouled out and may be the problem
- If the spark plugs look white or light grey the mixture is lean
- The ideal color is a light brown or mid grey color

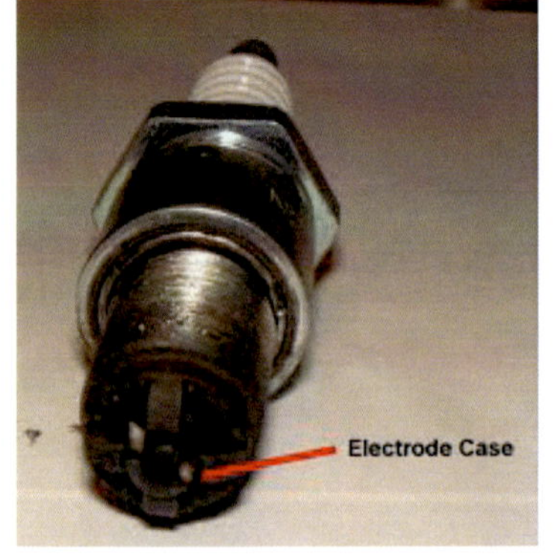

Chapter 4
Control and Diagnosis Systems

OBD I & OBD II Protocol

On-board diagnostics (**OBD**) is an <u>automotive</u> term referring to a vehicle's self-diagnostic and reporting capability. OBD systems give the vehicle owner or repair technician access to the status of the various vehicle subsystems

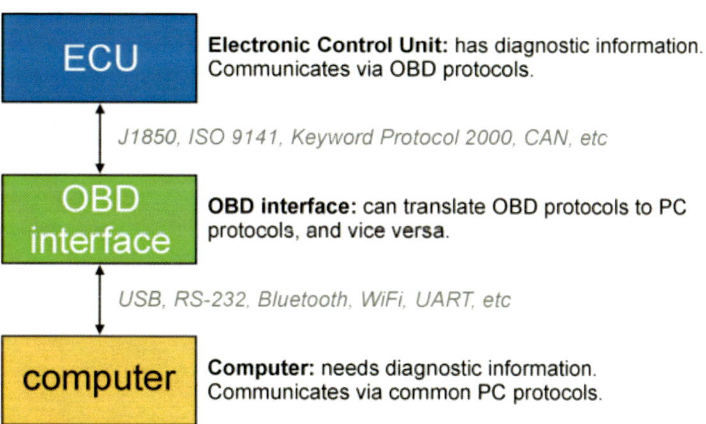

OBD I & OBD II Protocol

OBD systems are designed to monitor the performance of some of an engine's major components including those responsible for controlling emissions.

Learning how to work with those protocols also means that you can determine what that **Malfuction Indicator Light (MIL)** on your dash is referring to when it tells you there's an engine problem. If you or your mechanic has ever read the **DTCs (Diagnostic Trouble Codes)** on your engine, they are using OBD-II.

Malfunction Indicator Lamp (MIL)

The MIL is that terrible little light in the dash that indicates a problem with the system. There are a few variations, but they all indicate an error found by the OBD-II protocol

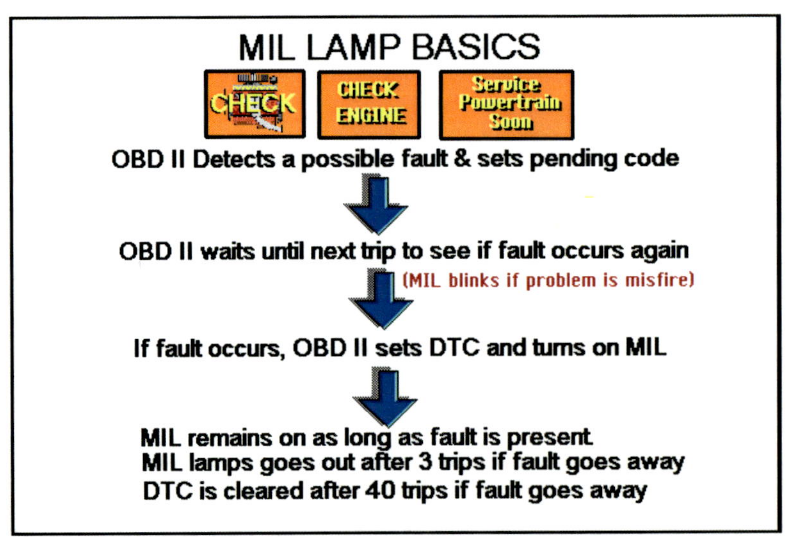

The Hardware

Any engine manufacture from 1996 or later is required by law to have the OBD-II computer system. You can access this system through the **Data Link Connector (DLC)**

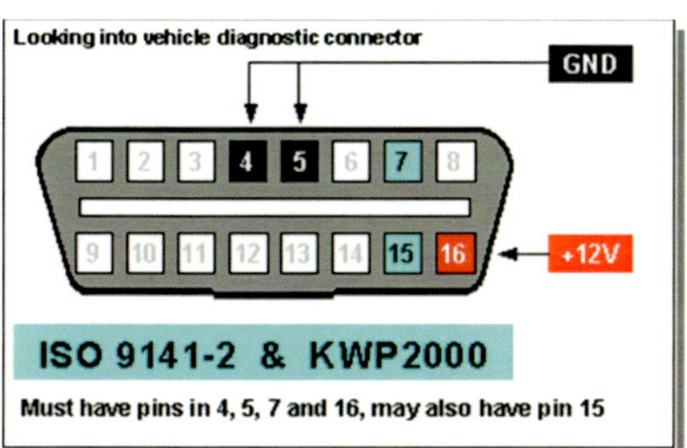

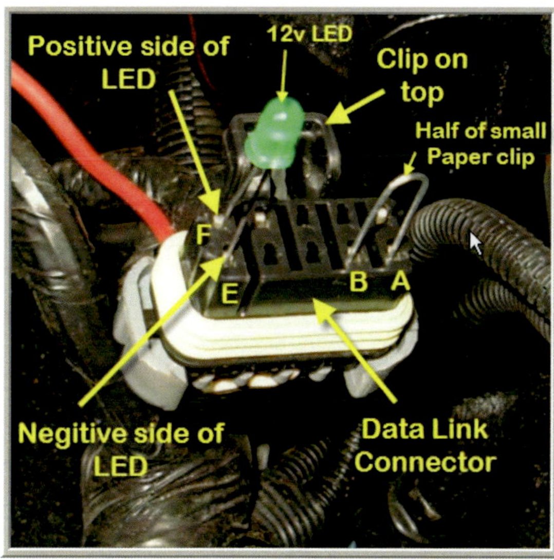

Diagnostic Trouble Code (DTC)

- These codes are used to describe where an issue is occurring on the vehicle and are defined by SAE (you can find the whole spec here for a cost). These codes, can either be generic or unique to the vehicle manufacturer.

- These codes take the following format:

- XXXXX

- First unit identifies the type of error code:
  - **P**xxxx for powertrain
  - **B**xxxx for body
  - **C**xxxx for chassis
  - **U**xxxx for class 2 network

Diagnostic Trouble Code (DTC)

- Second digit shows whether the code is manufacturer unique or not:
  - x**0**xxx for government-required code
  - x**1**xxx for manufacturer-specific code
- Third digit shows us what system the trouble code references
  - xx**1**xx/xx**2**xx show air and fuel measurements
  - xx**3**xx shows ignition system
  - xx**4**xx shows emissions systems
  - xx**5**xx references speed/idle control
  - xx**6**xx deals with computer systems
  - xx**7**xx/xx**8**xx involve the transmission
  - xx**9**xx notates input/output signals and controls

Diagnostic Trouble Code (DTC)

- Digits four and five show the specific failure code.
  - xxx**00** to xxx**99** - these are based on the systems defined in the third digit

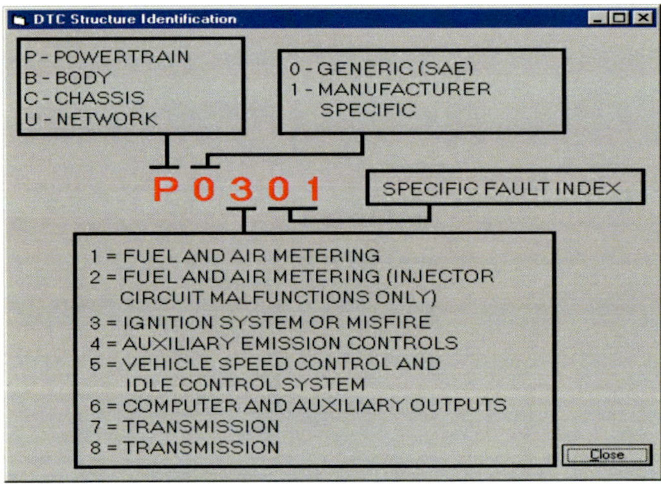

Marine Engines Protocols

- **Marine Diagnostic Trouble Code (MDTC or simply DTC)** . It is a marine protocol based on OBD II nomenclature
- The inboard and outboard gasoline engines uses their own protocol according with the **F**uel **I**njection Type.
    - MEFI 1
    - MEFI 2
    - MEFI 3
    - MEFI 4

Mercruiser GM Trouble Codes
MEFI - 1

- DTC Code 14 = Coolant Temp. Sensor
- DTC Code 21 = Throttle Position Sensor
- DTC Code 22 = Mass Air Flow
- DTC Code 23 = Manifold Air Temp.
- DTC Code 33 = Manifold Absolute Pressure
- DTC Code 42 = Electronic Spark Timing
- DTC Code 43 = Electronic Spark Control
- DTC Code 51 = Calibration Checksum Error

Mercruiser GM Trouble Codes
MEFI - 2

- DTC Code 14 = Coolant Sensor Voltage High (cold)
- DTC Code 15 = Coolant Sensor Voltage Low (hot)
- DTC Code 21 = Throttle Position Sensor Voltage High
- DTC Code 22 = Throttle Position Sensor Voltage Low
- DTC Code 23 = Manifold Air Temp Sensor High (cold)
- DTC Code 24 = Speed Sensor (if installed)
- DTC Code 25 = Manifold Air Temp Sensor Low (hot)
- DTC Code 33 = Manifold Absolute Pressure Sensor High
- DTC Code 34 = Manifold Absolute Pressure Sensor Low
- DTC Code 41 = Electronic Spark Timing Open Circuit
- DTC Code 42 = Electronic Spark Timing Grounded Circuit
- DTC Code 43 = Electronic Spark Control Detects Continuous Knock
- DTC Code 44 = Electronic Spark Control Cannot Detect Knock
- DTC Code 51 = ECM Calibration Checksum Error

Mercruiser GM Trouble Codes
MEFI – 3 & MEFI - 4

- DTC Code 13 = Oxygen Sensor 1 or 2 Malfunction
- DTC Code 14 = Coolant Sensor Voltage High (cold)
- DTC Code 15 = Coolant Sensor Voltage Low (hot)
- DTC Code 21 = Throttle Position Sensor Voltage High
- DTC Code 21 = Throttle Position Sensor Voltage Skewed High
- DTC Code 22 = Throttle Position Sensor Voltage Low
- DTC Code 23 = Manifold Air Temp Sensor High (cold)
- DTC Code 24 = Speed Sensor (if installed)
- DTC Code 25 = Manifold Air Temp Sensor Low (hot)
- DTC Code 31 = Governor Not Tracking
- DTC Code 32 = EGR Valve Not Tracking
- DTC Code 33 = Manifold Absolute Pressure Sensor High
- DTC Code 34 = Manifold Absolute Pressure Sensor Low
- DTC Code 41 = Electronic Spark Timing Open Circuit
- DTC Code 42 = Electronic Spark Timing Grounded Circuit
- DTC Code 43 = Electronic Spark Control Detects Continuous Knock
- DTC Code 44 = Electronic Spark Control Cannot Detect Knock
- DTC Code 51 = ECM Calibration Checksum Error

Mercruiser GM Trouble Codes
MEFI – 3 & MEFI - 4

- DTC Code 43 = Electronic Spark Control Detects Continuous Knock
- DTC Code 44 = Knock Sensor 1 or 2 Inactive
- DTC Code 51 = Calibration Checksum Error
- DTC Code 54 = Oxygen Sensor 1 or 2 Lean
- DTC Code 55 = Oxygen Sensor 1 or 2 Rich
- DTC Code 61 = Fuel Pressure Sensor High
- DTC Code 62 = Fuel Pressure Sensor Low
- DTC Code 63 = Fuel Temperature High
- DTC Code 64 = Fuel Temperature Low 81 See Scan Tool*

Mercruiser Data Link Connector

- The data link connector (DLC) is a 10-pin connector for attaching the diagnostic tool to the ECM
- The DLC is located on the port side of the engine next to the oil filter. Before attaching a diagnostic tool to the engine, verify that the key is off and the pins are clean of corrosion and debris

Mercruiser Data Link Connector

Locate the engine's 10 pin marine diagnostic connector (DLC). The location of this connector is specified in the engine manufacturer's service manual. Please be aware that the connector may have a protective cap attached to it

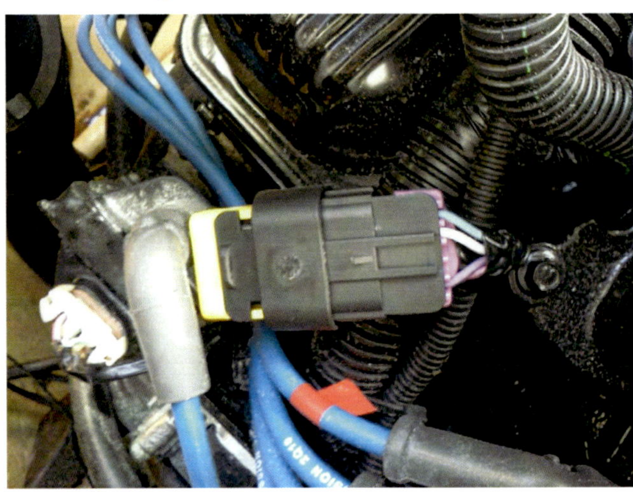

Diagnostic Tools

Diagnostic tools can only receive data with the key on or engine operating. Diagnostic tools need a minimum of 9.5 volts. If the diagnostic tool does not respond, verify the connection, verify that the key is on, and check the battery voltage

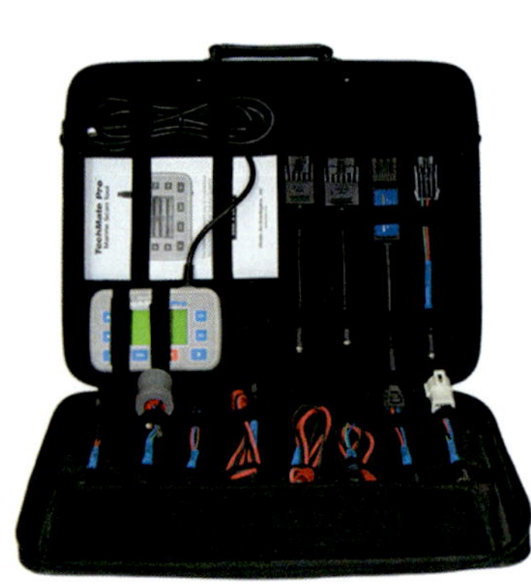

Electronic Control Unit
(ECU/ECM/PCM)

The ECU or ECM can refer to a single module or a collection of modules. These are the brains of the engine. They monitor and control many functions of the engine and transmission. These can be standard from the manufacturer, reprogrammable, or have the capability of being daisy-chained for multiple features

The ECM /PCM

The Electronic Module Control or Power Module control is 32 pin computer that manage the interaction between the sensors with some electromechanical devices such as , alternator, starter, governor, waste gate, EGR valves etc.

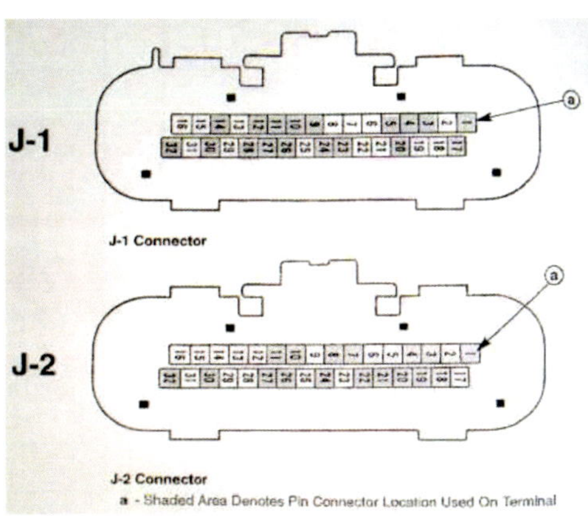

ECM Types of Connector

There are two types of connectors: J1 and J-2 , both of them with 32 pins and the following distribution

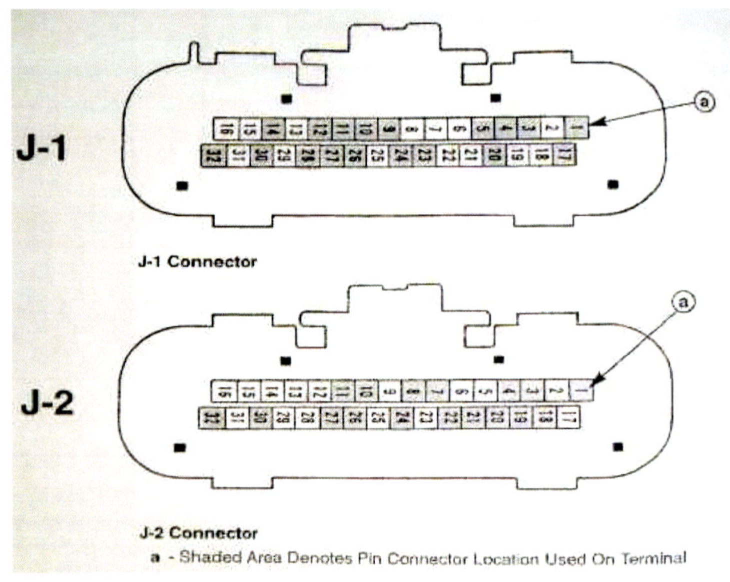

Connector J-1 Type

Pin No	Color	Description
1	LT Blue	Knock Sensor signal
2	WHT/BLK	DIAGNOSTIC .TEST. TERMINAL
3	Yellow	MASTER/SLAVE
4	TAN/BLK	GENERAL WARNING 2 / GOVERNOR MODE
5		EMERGENCY STOP
6	DK GRN/WHT	FUEL PUMP RELAY CONTROL
7		RPM CHANGE STATE
8	BRN/WHT	BUZZER
9		CHECK GAUGES / GOVERNOR OVERSPEED LAMP
10		
11	DK GRN	FUEL INJECTOR B DRIVER
12		EXHAUST GAS RECIRCULATION (EGR) VALVE
13	BLK	Ground

Connector J-1 Type

Pin No	Color	Description
14	Gray	Tachometer Out
15	LT GRN/BLK	IDLE AIR CONTROL (IAC) COIL .B. LOW
16	BLU/WHT	IDLE AIR CONTROL (IAC) COIL .A. HIGH
17	DK BLU	KNOCK SENSOR 1 SIGNAL
18		OIL LEVEL
19	TAN/BLK	GENERAL WARNING 1 / FCAL
20	YEL/BLK	LOAD 1 / LOCK LO / SHIFT INTERRUPT
21		LOAD 2 / TWIN ENGINE SYNC / TROLL MODE
22		GENERAL WARNING 1 LAMP / TRANS LAMP
23		GENERAL WARNING 2 LAMP / TROLL LAMP / GOVERNOR
24		OIL LEVEL LAMP
25		
26	DK BLU	FUEL INJECTOR A DRIVER

Connector J-1 Type

Pin No	Color	Description
27	BRN/WHT	CHECK ENGINE
28	Black	Ground
29	Black	Ground
30	LT GRN/WHT	IDLE AIR CONTROL (IAC) COIL .B. HIGH
31	BLU/BLK	IDLE AIR CONTROL (IAC) COIL .A. LOW
32		FUEL PRESSURE RELAY SENSE

Connector J-2 Type

Pin No	Color	Description
1	Orange	Battery Feed
2	Gray	5 Volt reference
3	BLK/WHT	SENSOR GROUND
4		FUEL PRESSURE SENSOR SIGNAL
5		EXHAUST GAS RECIRCULATION (EGR) VALVE FEEDBACK
6		HEATED OXYGEN (HO2) SENSOR .1. SIGNAL
7	Yellow	ECT SENSOR SIGNAL
8	LT GRN	MAP SENSOR SIGNAL
9		
10	ORN/BLK	SERIAL DATA
11		TAC SERIAL DATA 2
12	LT BLU	IGNITION CONTROL H
13	DK GRN	IGNITION CONTROL F

Connector J-2 Type

Pin No	Color	Description
14	PPL/WHT	IGNITION CONTROL D
15	TAN/BLK	IGNITION BYPASS
16	PPL/WHT	DISTRIBUTOR REFERENCE .HIGH.
17	RED/BLK	DEPSPOWER
18	YEL/BLK	DEPSLO
19	PNK/BLK	IGNITION FEED
20		OIL PRESSURE INPUT
21	TAN	IAT SENSOR SIGNAL
22		HEATED OXYGEN (HO2) SENSOR .2. SIGNAL
23	DK BLU TP	TP SENSOR SIGNAL
24		
25		SPEED SENSOR SIGNAL (VF)
26		SPEED SENSOR SIGNAL (AN) / FUEL TEMPERATURE

Connector J-2 Type

Pin No	Color	Description
27		TAC SERIAL DATA 1
28		
29		
30		
31	WHT	IGNITION CONTROL
32	BRN/WHT	CAM SENSOR SIGNAL

Scanner Connection

- With the engine's ignition switch in the OFF position, plug the scan tool's communication cable into the diagnostic connector

- Once the tool is connected, turn the ignition switch ON and start the engine if necessary. If the engine will not start, simply leave the ignition switch ON and proceed

Scanner Connection

After the scan tool displays its initial opening messages, use the Up and Down keys to select the "Marine EFI" operating mode, then press the YES key to select it

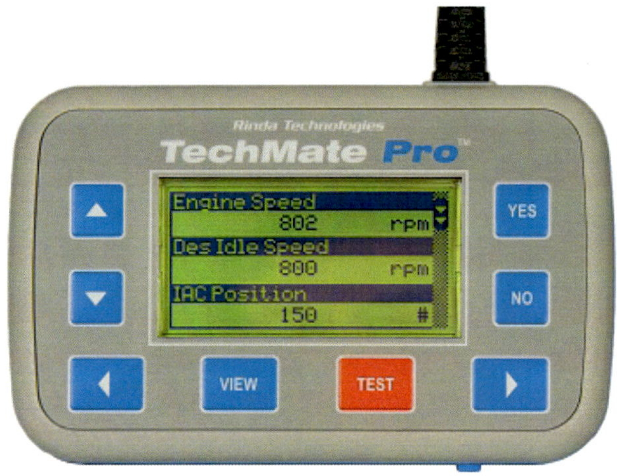

The Wiring Harness

The wiring harness connects the primary fuse box, or all fuse boxes on some twin engines models with the ECM and each sensor or electrical device. The fuse box takes power from the ignition switch during the cranking and then from the alternator when the engine is running

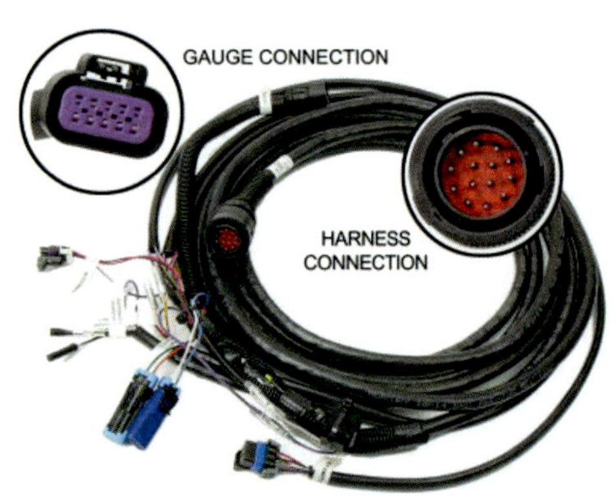

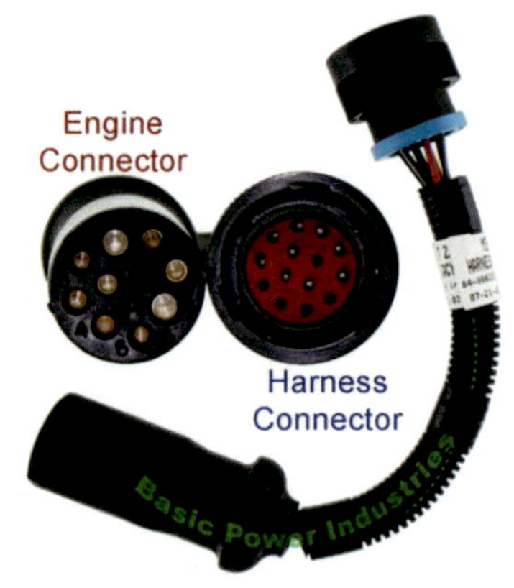

The Wiring Harness

Components commonly connected to the wiring harness include the on-board diagnostic computer (ECM) , engine sensors , transmission sensors , fuel-injection system , auxiliary lights and dashboard controls. The emissions system , alternator , starter and the fire suppression system are also connect directly to the harness

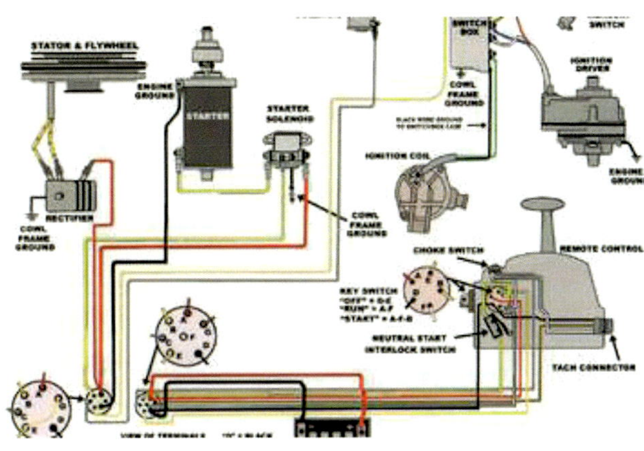

Engine Sensors

- It is first important to understand that an OBDII code or DTC code in itself is not indicative of a sensor failure. Sensors simply report information

- No matter how complicated, you'll find the sensors below in any EFI engine:

  - » Engine coolant temperature - CTS
  - » Air temperature
  - » Barometric pressure/manifold absolute pressure - MAP
  - » Mass air flow - MAF
  - » Idle air controller - IAC
  - » Crankshaft
  - » Camshaft
  - » Throttle position - TPS
  - » Oxygen – O2
  - » Knock

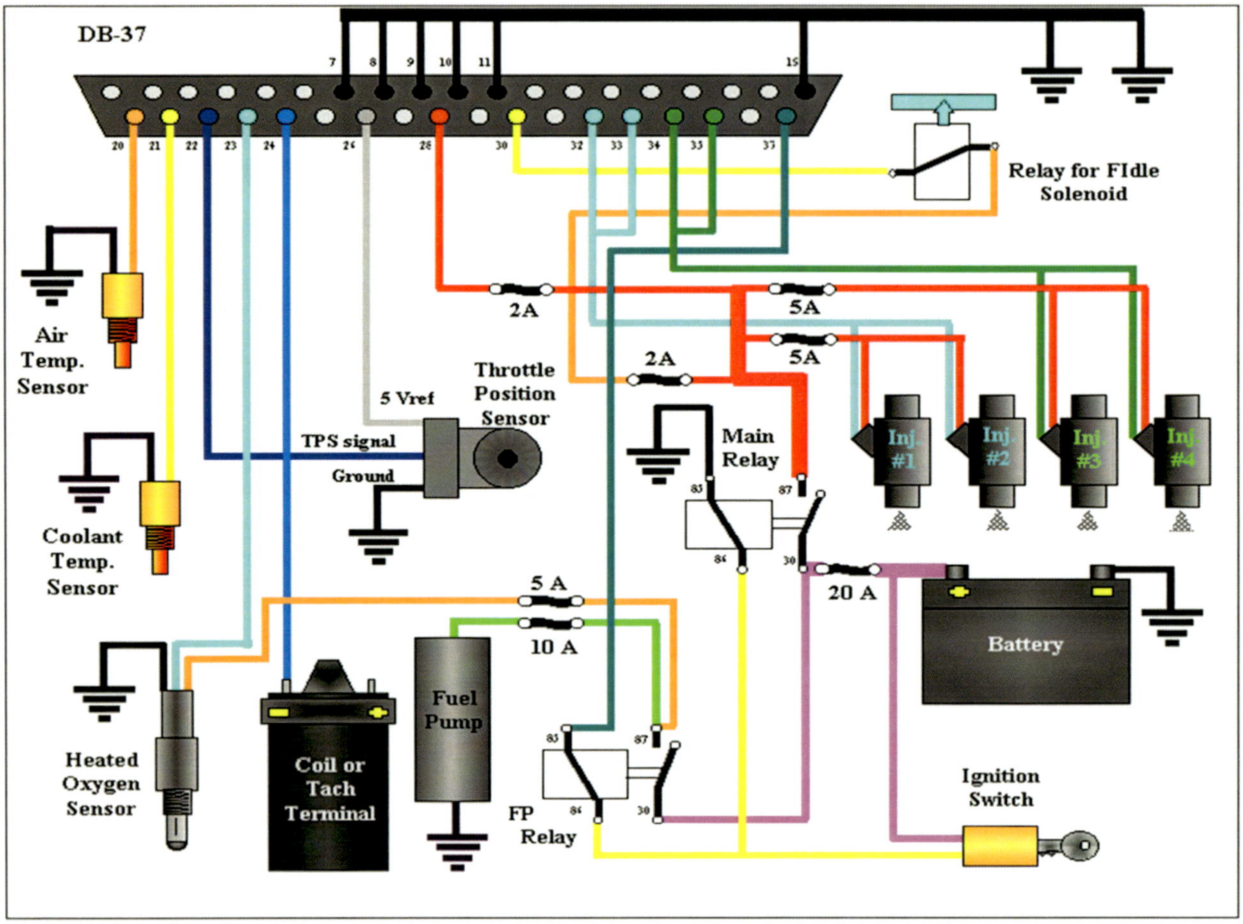

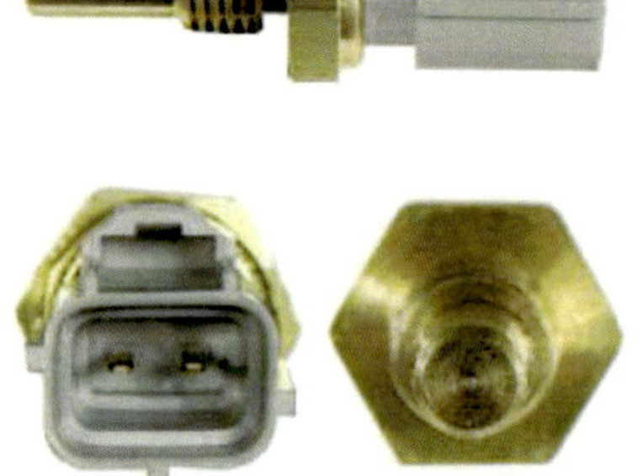

Coolant Temp Sensor CTS

The engine coolant temperature sensor simply tells the ECM the current temperature of the engine. When the temperature of the coolant reaches between 75 and 95 degrees Celsius (depending on the manufacturer specifications), the thermostat opens and start the coolant circulation through the heat exchanger.

Coolant Temp Sensor CTS

Usually it is located at the thermostat housing , sharing place with the <u>Coolant Temp Switch</u>. Depending of the amount of Temp gauges are the sensors. The recommendation is one **sensor** per each **temp gauge**

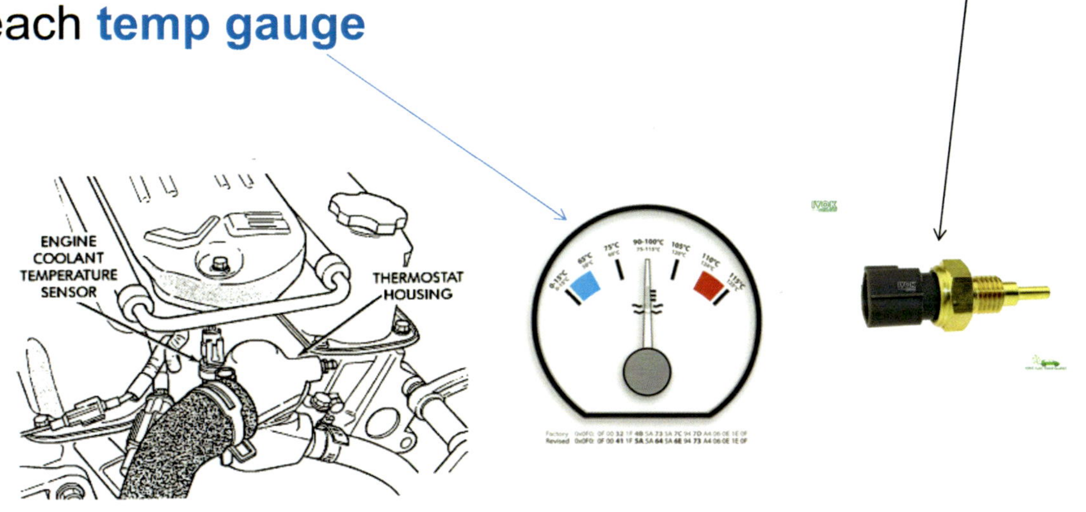

CTS Servicing

- It can be cleaned using a wire brush when you entirely change your engine coolant (approx. every 500 Hrs)

- Also the plug and the terminals should be cleaned periodically

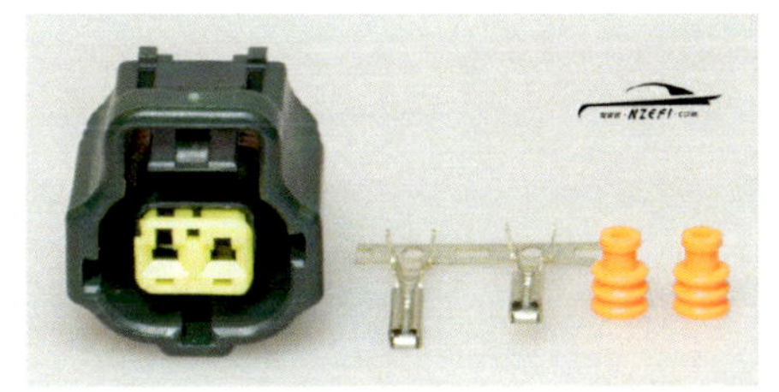

CTS Sensor Diagnosis

- The ECM supplies 5 volts to the ECT sensor circuit. The sensor is a thermistor which changes internal resistance as temperature changes
- When the sensor is cold (internal resistance high), the ECM monitors a high signal voltage and interprets it as a cold engine

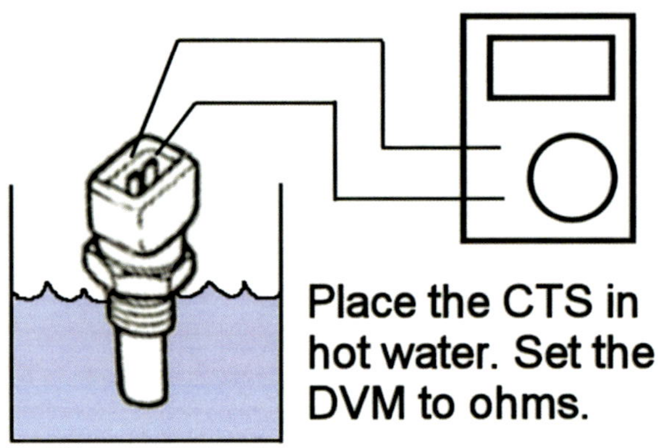

Place the CTS in hot water. Set the DVM to ohms.

CTS Sensor Diagnosis

As the sensor warms (internal resistance decreases), the voltage signal will decrease and the ECM will interpret the lower voltage as a warm engine

Temperature		Engine Coolant/Air Charge Temperature Sensor Values	
°F	°C	Voltage (volts)	Resistance (K ohms)
248	120	.27	1.18
230	110	.35	1.55
212	100	.46	2.07
194	90	.60	2.80
176	80	.78	3.84
158	70	1.02	5.37
140	60	1.33	7.70
122	50	1.70	10.97
104	40	2.13	16.15
86	30	2.60	24.27
68	20	3.07	37.30
50	10	3.51	58.75

CTS Sensor Readings

Temperature		Engine Coolant/Air Charge Temperature Sensor Values	
°F	°C	Voltage (volts)	Resistance (K ohms)
248	120	.27	1.18
230	110	.35	1.55
212	100	.46	2.07
194	90	.60	2.80
176	80	.78	3.84
158	70	1.02	5.37
140	60	1.33	7.70
122	50	1.70	10.97
104	40	2.13	16.15
86	30	2.60	24.27
68	20	3.07	37.30
50	10	3.51	58.75

Check Readings

- Check With Cold Sensor
  - Connect the black lead of the meter to the body of the cold sensor and the red to the terminal. You should have a reading of approximately 2000 ohms. Check the warm sensor in your engine. You should see a much lower reading on the digital meter. If not, then it is not working correctly
- Check With Hot Sensor
  - While the engine is running checking with the digital ohm meter. You are looking for readings that are more than 200 ohms in variance between a cold and warm engine. If you do not see anything that is more than 200 ohms in difference, the coolant temperature sensor is defective and will need to be replaced.

CTS Sensor Codes

The OBD II codes related with a failure of an CTS sensor are :

- PO115 to PO118 ,P0128,P2185 (Generic Codes)
- DTC Codes: 14 , 15, (Mercruiser Scanner Tool)

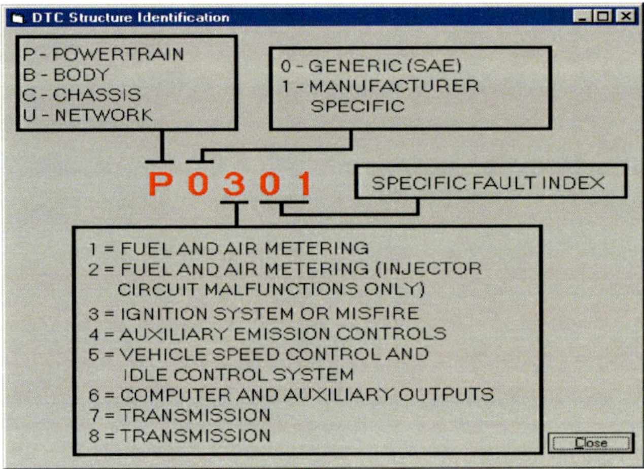

Intake Air Temp Sensor (IAT)

The **Intake Air Temperature sensor (IAT)** monitors the temperature of the air entering the engine. The ECM needs this information to estimate air density so it can balance air air/fuel mixture

Intake Air Temp Sensor (IAT)

The ECM converts the resistance of the intake air temperature sensor to degrees. Intake Air Temperature (IAT) is used by the ECM to adjust fuel delivery and spark timing according to incoming air density.

Temperature °F	Voltage
-40°F	4.90 V
+33°F	4.75 V
+68°F	4.00 V
+100°F	3.00 V
+143°F	2.00 V
+176°F	1.30 V
+248°F	0.60 V
+305°F	0.0 V

IAT Sensor Location

It is usually located on the air filter box or the pipe going from the air filter box to the throttle body. Some manufacturers have a MAF and an IAT sensor built in one unit which has five wires located on the air filter box

How The IAT Sensor Works

It works the same as a coolant sensor. The air temperature sensor is a thermistor, which means its electrical resistance changes in response to changes in temperature

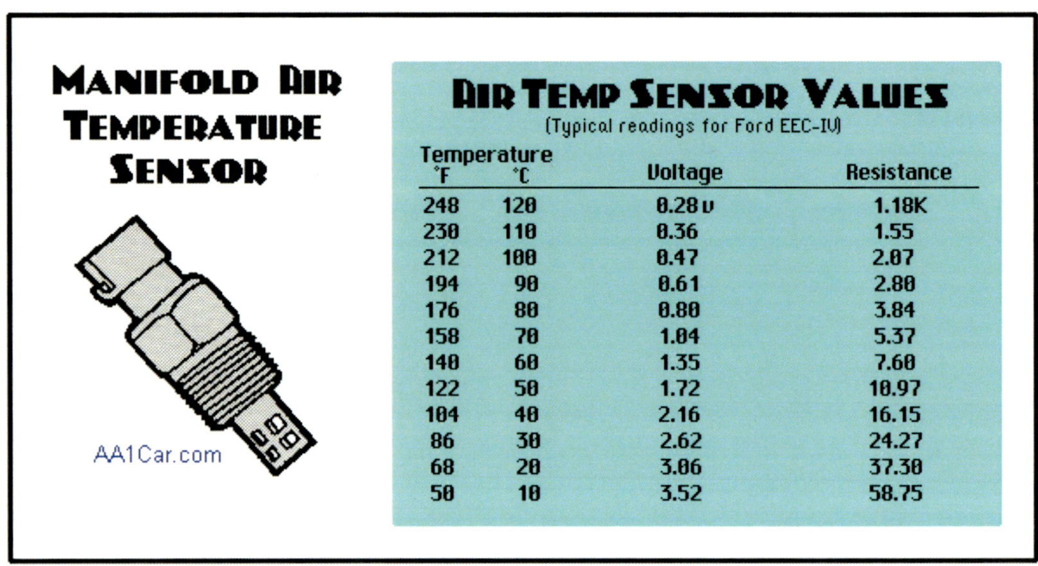

AIR TEMP SENSOR VALUES
(Typical readings for Ford EEC-IV)

Temperature °F	°C	Voltage	Resistance
248	120	0.28 v	1.18K
230	110	0.36	1.55
212	100	0.47	2.07
194	90	0.61	2.80
176	80	0.80	3.84
158	70	1.04	5.37
140	60	1.35	7.60
122	50	1.72	10.97
104	40	2.16	16.15
86	30	2.62	24.27
68	20	3.06	37.30
50	10	3.52	58.75

How The IAT Sensor Works

The ECM applies a reference voltage to the sensor (usually 5 volts), then looks at the voltage signal it receives back to calculate air temperature. The return voltage signal will change in proportion to changes in air temperature

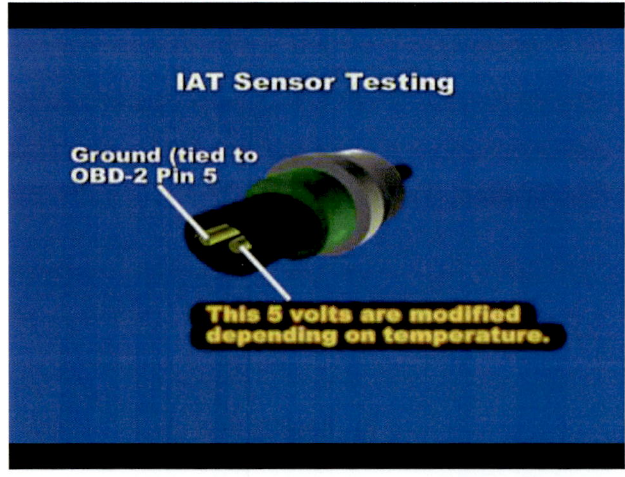

Symptoms of a bad IAT Sensor

- **Lack of power when accelerating**
  - A bad or failing sensor will not send the correct signal to the computer which will negatively affect the computer's air / fuel mixture calculations.

- **Trouble with cold starts**
  - Another symptom of a bad or defective air charge temperature sensor is difficulty starting in cold conditions. The air charge temperature sensor sends a signal to the computer during cold conditions that allow the computer to provide additional fuel required during cold starts. A failing or defective sensor will not be able to send the correct signal to the ECU (engine control unit), which can result in increased difficulty when cold starting a boat

IAT Sensor Codes

The OBD II codes related with a failure of an IAT sensor are :

- PO110 to PO113 (Generic Codes)
- DTC Code: 23 (Mercruiser Scanner Tool)

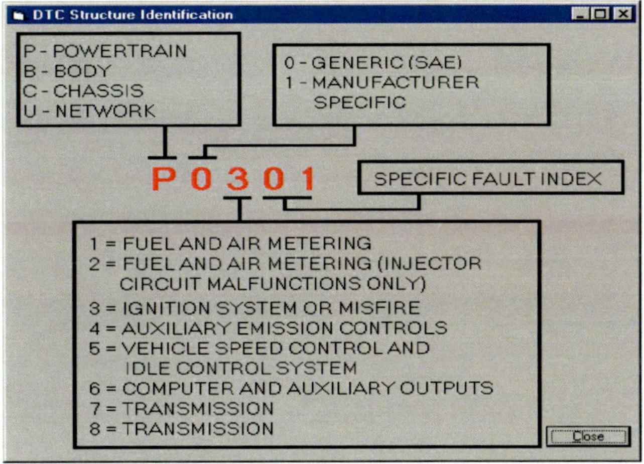

Manifold Absolute Pressure MAP

The Manifold Absolute Pressure (MAP) sensor measures the change in the intake manifold pressure from engine load and speed changes

	atmospheric pressure at sea level					approximate engine idle condition					absolute zero pressure
BAR	1.0	0.9	0.8	0.7	0.6	0.5	0.4	0.3	0.2	0.1	0.0
kPa	100	90	80	70	60	50	40	30	20	10	0
approx MAP voltage	4.9	4.4	3.8	3.3	2.7	2.2	1.7	1.1	0.6	0.3	0.3

Manifold Absolute Pressure MAP

As intake manifold pressure increases, intake vacuum decreases resulting in a higher MAP sensor voltage and kPa reading.

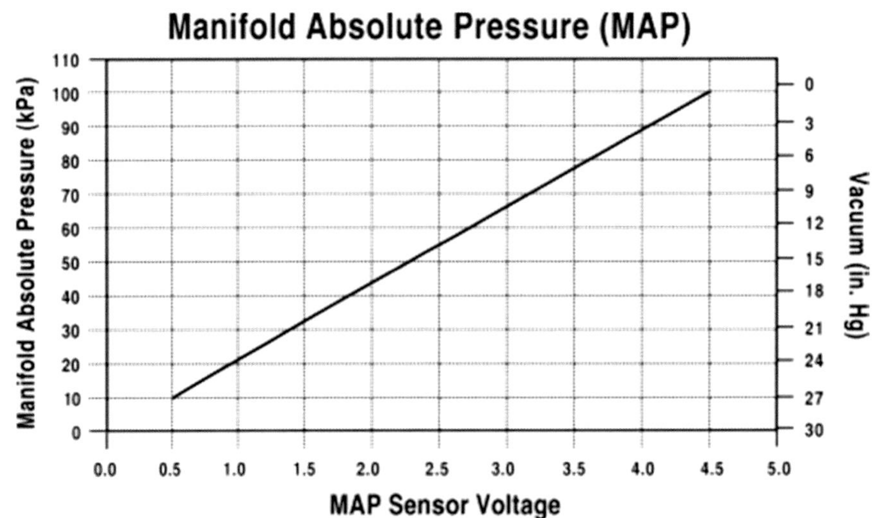

Manifold Absolute Pressure MAP

When the engine is off, the absolute pressure inside the intake equals atmospheric pressure, so the MAP will indicate about 14.7 psi. At a perfect vacuum, the MAP sensor will read 0 psi.

Vacuum at sea level (in. Hg.)	Manifold Absolute Press. (kPa)	Sensor Voltage
0	101.3	4.50
3	91.2	4.00
6	81.0	3.50
9	70.8	3.00
12	60.7	2.50
15	50.5	2.00
18	40.4	1.50
21	30.2	1.00
24	20.1	0.50

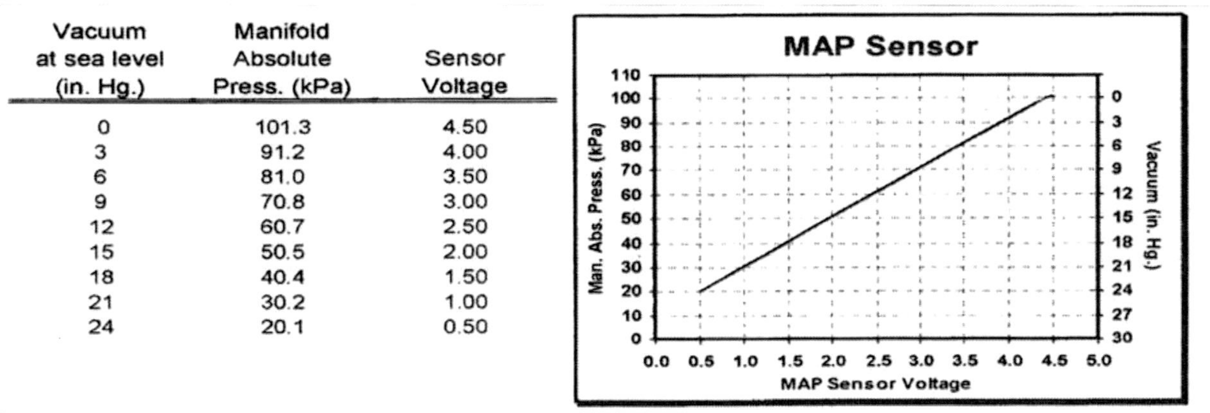

MAP Sensor - Vacuum

When the engine is running, the downward motion of the pistons create a vacuum inside the intake manifold. (For the purposes of engine control, when we say vacuum, what we are really saying is pressure that is less than atmospheric pressure . With a running engine, intake manifold vacuum usually runs around 18 - 20 "Hg (inches of mercury). At 20 "Hg, the MAP sensor will indicate about 5 psi.

Note: **I Atm = I4.7 PSI = 30 in of Hg**

MAP Sensor Location

Usually found either bolted on the intake manifold or linked with a vacuum pipe from the intake manifold

TYPICAL LOCATION

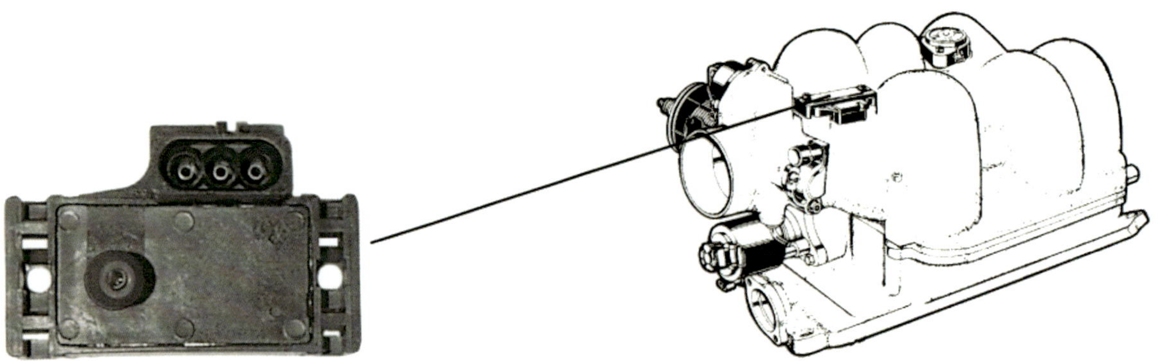

MAP Sensor Servicing

If mounted on intake manifold then should be cleaned every 6 months to 1 year, using carb cleaner (depends on fuel quality where you live; the lower the quality of fuel, the more carbon deposits found). I've had to clean some every three months

MAP Sensor Volt / Pressure

A Manifold Absolute Pressure (MAP) sensor lets the ECU know how much pressure is in the intake manifold at any given time

Altitude	Baro Pressure	Voltage
0 to 1,000'	29-30"	4.5-4.8 V
1,000-2,000'	28-29"	4.3-4.5V
2,000-3,000'	27-28"	4.1-4.3 V
3,000-4,000'	26-27"	3.9-4.1 V
4,000-5,000'	25-26"	3.8-4.0 V
5,000-6,000'	24-25"	3.7-3.9 V
6,000-7,000'	23-24"	3.6-3.8 V
7,000-8,000'	22-23"	3.4-3.6 V
8,000-9,000'	21-22"	3.3-3.5 V
9,000-10,000'	20-21"	3.1-3.3 V

MAP Sensor Testing

- Remove the MAP sensor from the intake manifold (if it's bolted directly onto the Intake Manifold).

- Connect your vacuum pump to the MAP sensor's vacuum nipple. You probably had to disconnect the MAP sensor from its electrical connector to remove it, so if you did.. reconnect it to it now

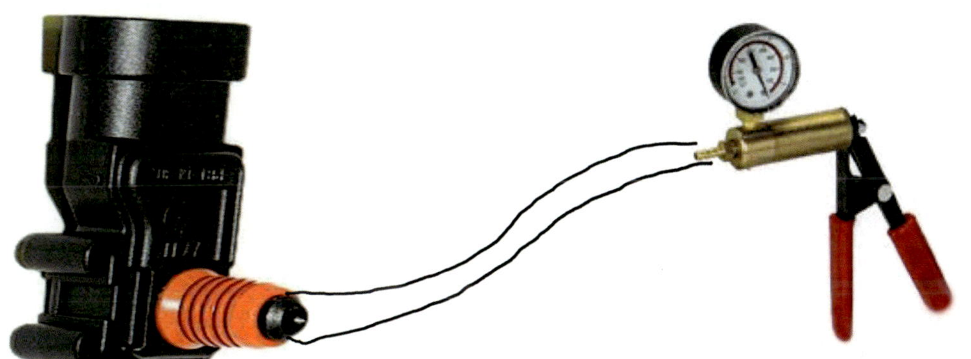

MAP Sensor Testing

- Set your multimeter's selector to Volts DC mode and probe the wire labeled with **number 2** (The middle one)

- Remember, the MAP sensor must remain connected to its 3 wire connector

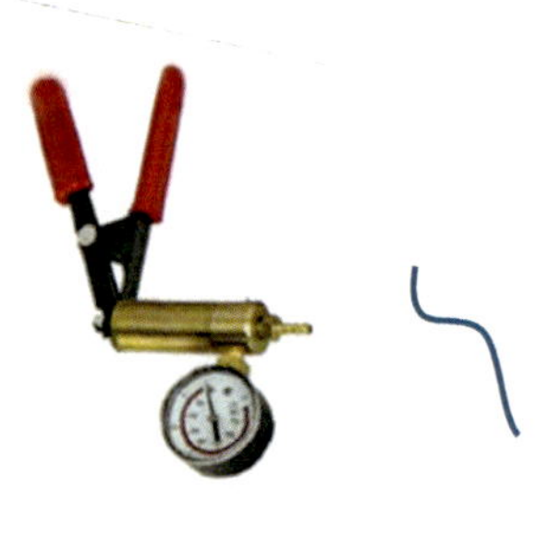

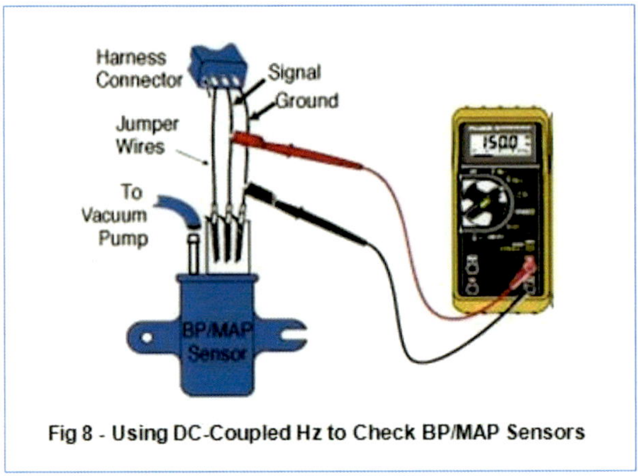

Fig 8 - Using DC-Coupled Hz to Check BP/MAP Sensors

MAP Sensor Testing

When everything is ready, turn the Key on but don't start the engine. This will power up the MAP sensor and you should see a reading of 4.7 Volts DC on your multimeter

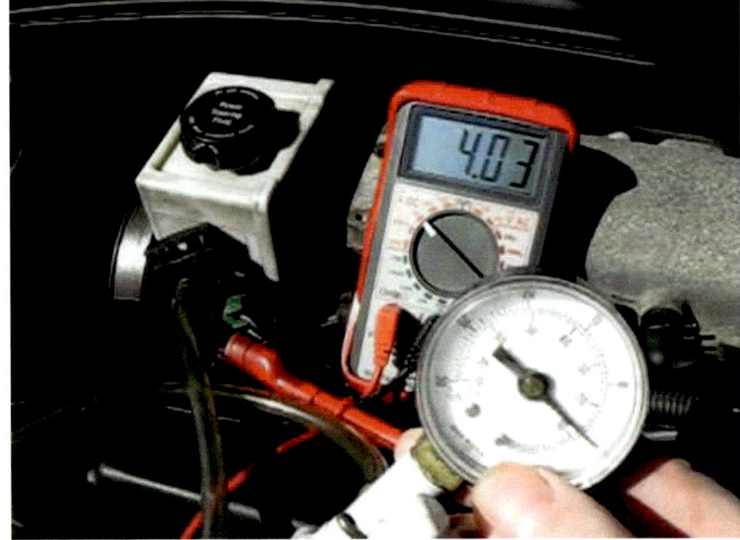

MAP Sensor Testing

Now, apply vacuum to the MAP sensor with the vacuum pump (or your mouth). The multimeter should display the following DC voltages at the following vacuum values if your using a vacuum pump:

1.) 0 in. Hg ...... 4.7 Volts.

2.) 5 in. Hg ...... 3.9 Volts.

3.) 10 in. Hg .... 3.0 Volts.

4.) 20 in. Hg .... 1.1 Volts

MAP Sensor Testing

During the vacuum application, you should see a smooth increase and decrease (without any gaps or skips) on the multimeter. You should see the voltage decrease and increase as you apply and release vacuum.

MAP Sensor Testing Results

- **The multimeter registered the indicated voltages as you applied vacuum**
  - This means that the (MAP) sensor is good and not the cause of the MAP sensor code or problem on your engine
- **The multimeter registered voltage, BUT it did not increase or decrease as you applied vacuum**
  - This confirms that the (MAP) sensor on your engine is BAD. Replacing the MAP sensor will solve the MAP code issue (P0106, P0107, P0108 or 33, 34)
- **The multimeter registered 0 Volts**
  - This usually means that the MAP sensor is fried. To be absolutely sure, I suggest confirming that the MAP sensor has power and ground. If both (power and ground) are present, the MAP is BAD

Symptoms of a bad MAP Sensor

- **Excessive fuel consumption**
  - A MAP sensor that measures high intake manifold pressure indicates high engine load to the PCM. This results in an increase of fuel being injected into the engine
- **Lack of power**
  - A MAP sensor that measures low intake manifold pressure indicates low engine load to the PCM. The PCM responds by reducing the amount of fuel being injected into the engine. While you may notice an increase in fuel economy, you will also notice that your engine isn't as powerful as it was before
- **Failed emissions test**
  - Your tailpipe emissions may show a high level of hydrocarbons, high NOx production, low CO2, or a high level of carbon monoxide.

MAP Sensor Codes

The OBD II codes related with a failure of an MAP sensor are :

- P1105 to P1108 and P0342 (Generic Codes)
- DTC Code 33,34,36 & 37 (Mercruiser Scanner Tool)

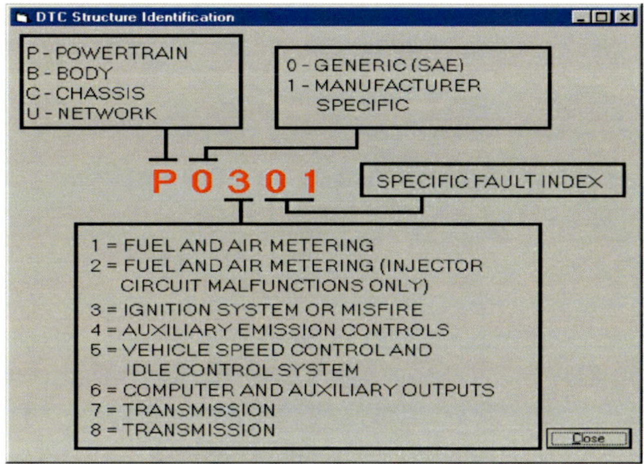

Mass Air Flow Sensor MAF

The Mass Air Flow Sensor is probably the best way to measure the amount of air an engine takes in (engine load). This sensor not only measures the volume of air but also compensates for its density as well

How MAF Sensor Works

A resistance-calibrated wire is mounted in the engine's intake airflow before the throttle and supplied with a constant voltage. The sensor's electronics control the current flowing through the wire to heat it to a specific temperature. Air flowing past the wire carries away heat, so the current must be increased to maintain its temperature

How MAF Sensor Works

A typical analog sensor has two elements that measure temperature, a "cold" wire that measures the intake air temperature, and a "hot" wire mounted next to it that is maintained at a specific temperature above the cold wire

MAF Sensor Measurements

Electronics inside the sensor reads both temperatures, controls the level of current flowing through the hot wire and produces a 0 – 5-volt output signal that's proportional to the current applied to the hot wire. The output signal is typically about **0.7 volts at idle** and about **4.5 volts at wide-open throttle**

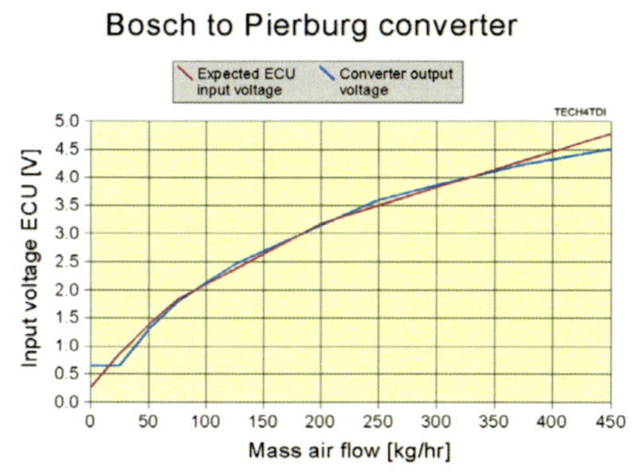

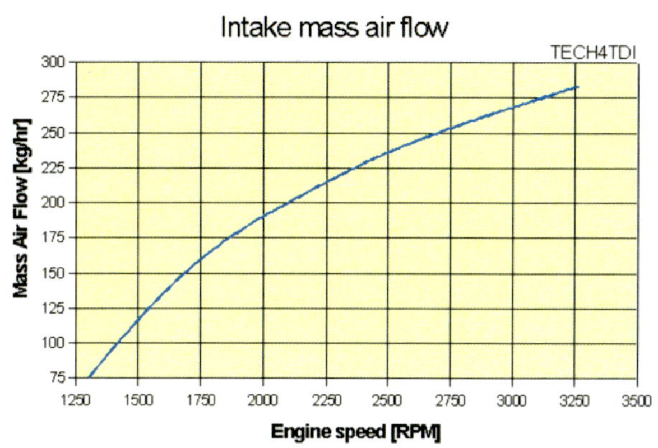

Mass Air Flow Sensor MAF

There are two common designs of MAF sensors used in today's vehicles. One produces a variable voltage output (analog) and the other produces a frequency output (digital). In either case their operation is similar. Both outputs can be measured by a scanner or a digital volt/ohm meter that can measure frequency

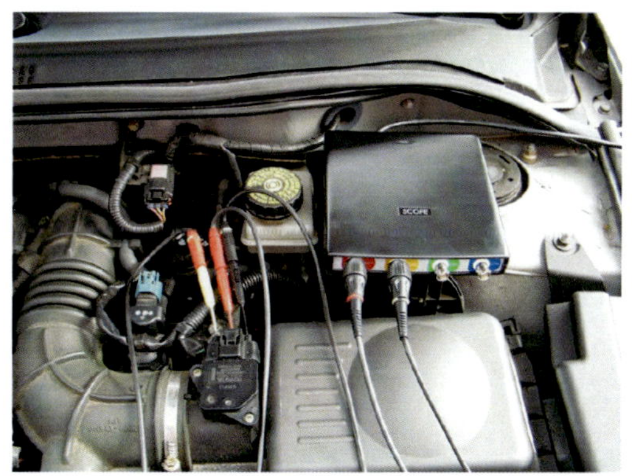

Hot Wire Principle

Both designs work on the "hot wire" principle. Here's how they work. A constant voltage is applied to the heated film or heated wire. This film or wire is positioned in the air stream or in an air flow sampling channel and is heated by the electrical current that the voltage produces

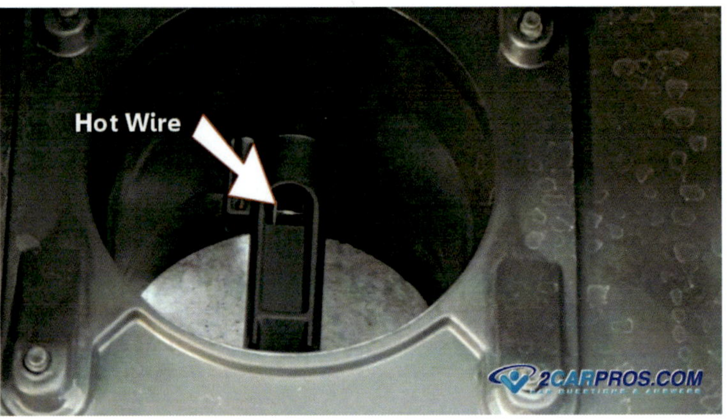

Hot Wire Principle

As air flows across it, it cools down. The heated wire or film is a positive temperature coefficient (ptc) resistor. This means that it's resistance drops when it's temperature drops. The drop in resistance allows more current to flow through it in order to maintain the programmed temperature

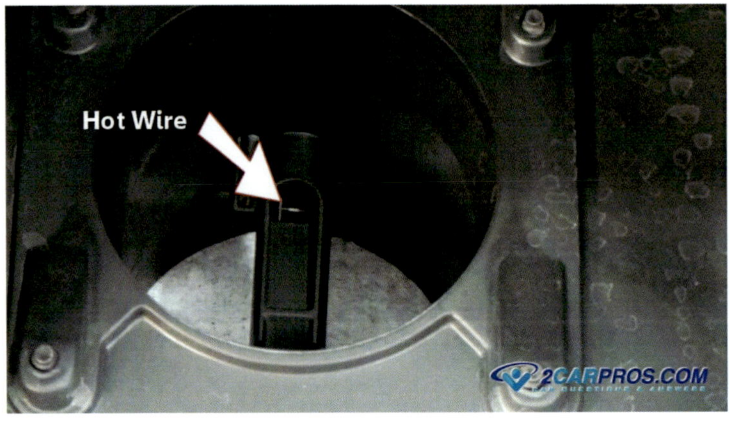

Hot Wire Temperature

- On some sensors, the hot wire is momentarily heated to about 1,000 degrees F when the ignition is turned off to burn it clean.

- Usually with a dirty MAF, long-term fuel trim will be negative at idle and become more positive as rpm (airflow) increases

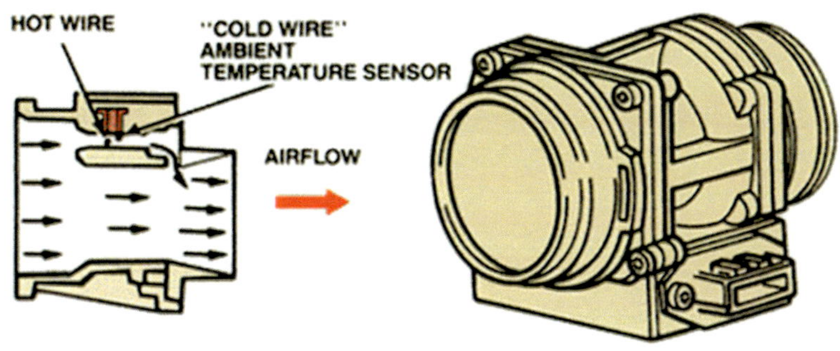

MAF Sensor / Humidity

Air temperature affects density since colder air is more dense than warmer air. Many systems use an air temperature sensor to compensate for this factor since similar amounts of air can enter an engine at different temperatures

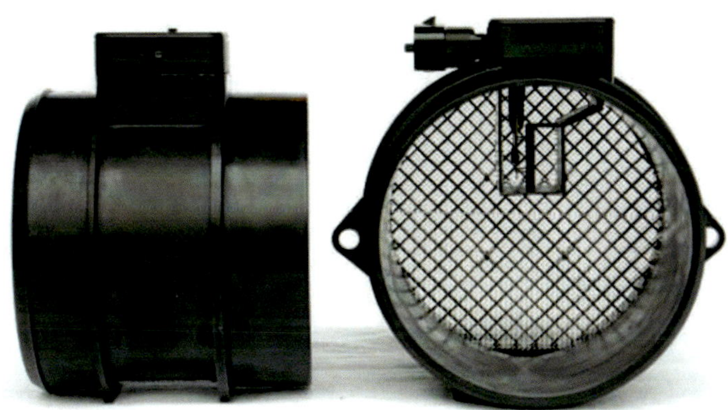

MAF Sensor / Air Filter

By far the most common MAF failure is caused by accumulation of dirt on the hot wire. Even though the sensor is protected by the air filter (it's often mounted inside the air filter housing), oil mist or fine dust that contacts the hot wire tends to bake onto the wire

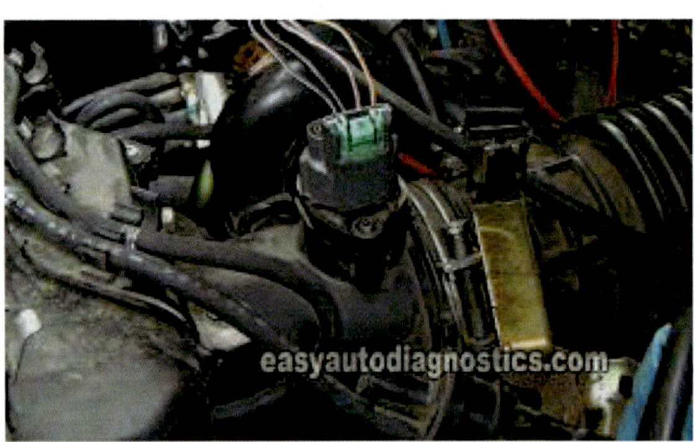

MAF Sensor Service

- Most OEM service information will tell you a MAF cannot be cleaned. Sometimes this is true because the hot and cold wires (which look like tiny resistors) simply cannot be accessed

- Cleaning a MAF sensor is a good way to confirm your diagnosis before spending the customer's money. If the MAF is dirty enough the cause the driveability problem, even partial cleaning will restore some of the engine's performance

- There are special chemicals available for cleaning a MAF sensor, but many techs are comfortable using any spray cleaner that dries quickly and doesn't leave a residue

MAF and IAT Sensors

On many models, the MAF sensor also reports intake air temperature (IAT). On some the PCM simply uses the signal from the "cold" wire for IAT. Other MAF sensors include a separate (non-replaceable) IAT sensor as part of the same assembly

That IAT sensor has its own reference voltage and signal return wires, so these MAF sensors have at least five wires

MAF Sensor Codes

The OBD II codes related with a failure of a MAF sensor are :

- P0100 to P0104 P0171 ,P0174 (Generic Codes)
- DTC Codes: 22 (Mercruiser Scanner Tool)

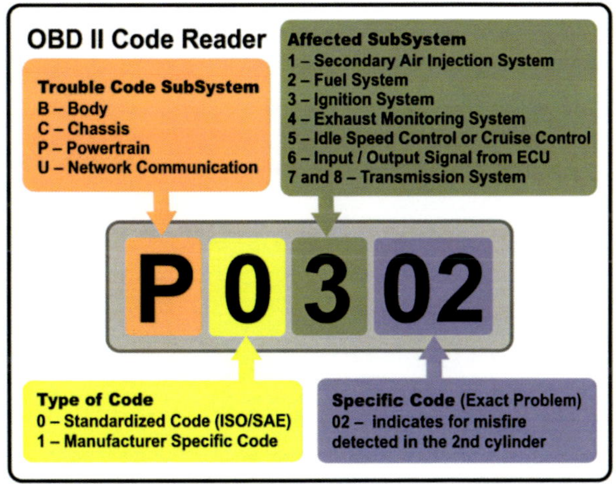

Crankshaft Position Sensor

The [crankshaft position sensor](#) (CKP sensor or CPS) is one of several sensors that keep your engine running smoothly. It measures the position of the crankshaft (for which it is sometimes called the crank angle sensor or CAS)

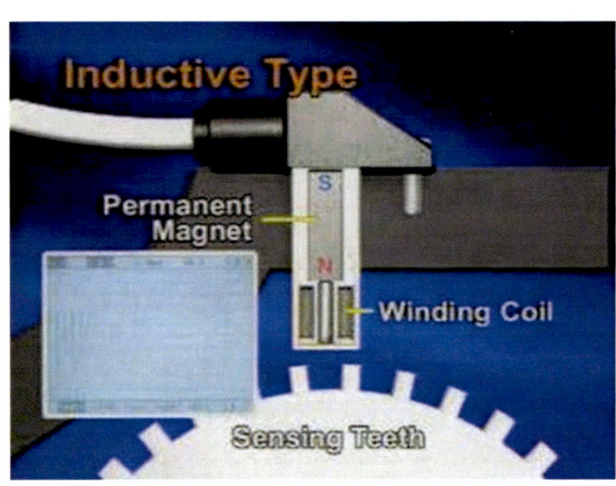

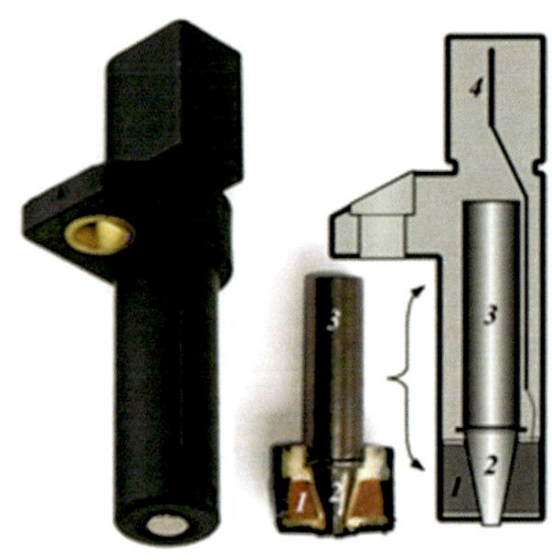

How a CKP Sensor Work

There are different setups for crankshaft sensors, but they tend to work on principles of magnetism. Many crankshaft position sensors are a type of electronic sensor known as a Hall effect sensor. A Hall effect sensor produces electricity in when it is exposed to a magnetic field

In a crankshaft position sensor, a toothed wheel, spinning with the crankshaft, disrupts the magnetic field. That produces a pattern of on and off switches in the Hall Sensor that the ECU can interpret as the crankshaft speed. The faster the sensor is turning on and off, the faster the crankshaft is spinning

CKP Sensor / RPM

The sensor produces an electrical voltage based on fluctuations in a magnetic field. The fluctuations are caused by the movement of metal pins in the crankshaft. Faster spinning means more fluctuations, means more voltage. The ECU is able to translate this voltage into crankshaft speed

CKP Sensor Location

The location of the crankshaft position sensor can vary from one vehicle to another. Obviously it must be close to the crankshaft, so it is most often located on the front underside of the engine. It can usually be found mounted to the timing cover. Sometimes it may be mounted at the rear or the side of the engine

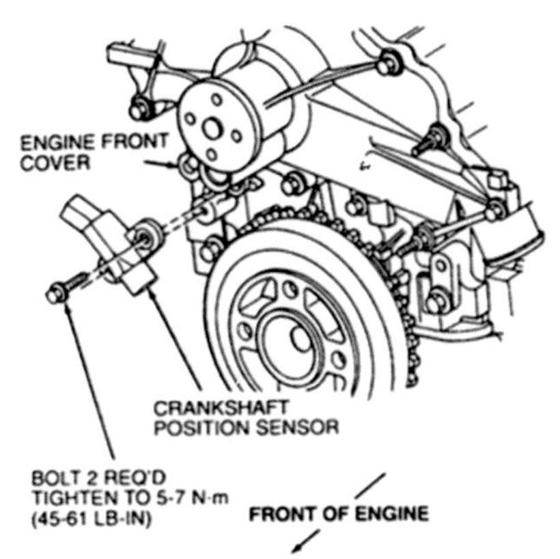

Symptoms of a Bad CKP Sensor

- **Cylinder Misfiring**
  - Only one of the cylinders needs to misfire to cause a problem. Many times misfiring is due to issues with the spark plug, but if the sensor fails then the computer is not getting the correct information on piston position

- **Acceleration Hesitation**
  - This is also similar to the cylinder misfiring. If there is some type of hesitation in acceleration, then the computer is not getting the correct information on the position of the cylinders when the car is running

- **Intermittent Starting**
  - When the sensor completely fails then the internal computer will have a malfunction code. If the car continues to have difficulty starting and you do not have it fixed, then one day it will not start at all. Many times this type of failure is because there is a problem with the electrical connection

Symptoms of a Bad CKP Sensor

- **Engine Vibrations**
  - Once the sensor fails then the position is not monitored, so you may find that the engine starts to have very strong vibrations. This will in turn affect the engine power and can also stop the mileage from recording properly

- **Backfiring and Stalling**
  - Your engine may stall from time to time. It can stall a few seconds after starting or quite some time after starting, and the same goes for backfiring. You will also notice that the car keeps backfiring frequently, not just every so often

- **Irregular Engine Function**
  - Basically you will notice that regular engine workings will start to become irregular, such as speed fluctuations, irregular acceleration, idling fluctuations, and more

CKP Sensor Testing

- You can remove the sensor, and then test the resistance. Attach one end of the multimeter to each wiring lead of the sensor. Resistance of zero means that the there is a short circuit. Infinite resistance means there is an open circuit. Either one of those readings indicates that the sensor is not working

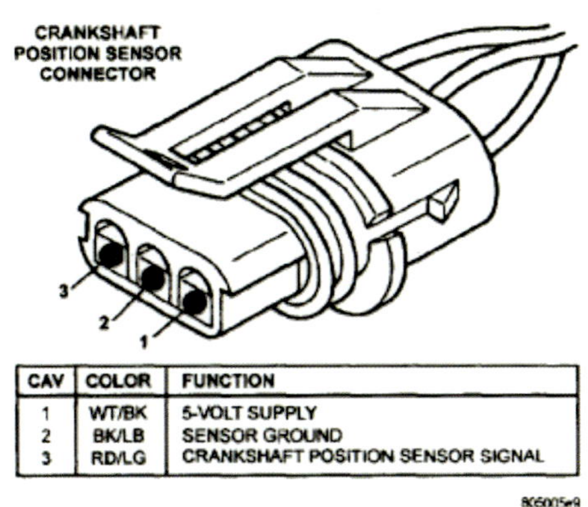

CAV	COLOR	FUNCTION
1	WT/BK	5-VOLT SUPPLY
2	BK/LB	SENSOR GROUND
3	RD/LG	CRANKSHAFT POSITION SENSOR SIGNAL

CKP Sensor Testing

Another way to test the crankshaft sensor with a multimeter is by checking the output voltage _with the engine cranking_. You will need an assistant to do this. Be very careful around moving parts as you do this. Probe the wiring connectors and measure the output voltage in AC millivolts. Typically, this reading is around **200 millivolts**

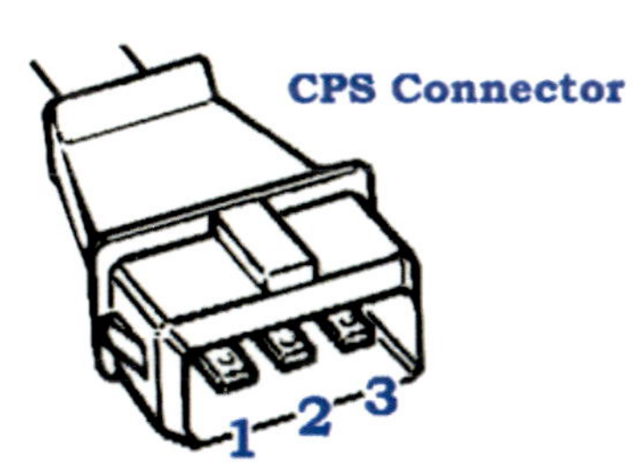

CKP Sensor Codes

The OBD II codes related with a failure of an CKP sensor are :

- PO335 to PO338 (Generic Codes)
- DTC Code: 24 (Mercruiser Scanner Tool)

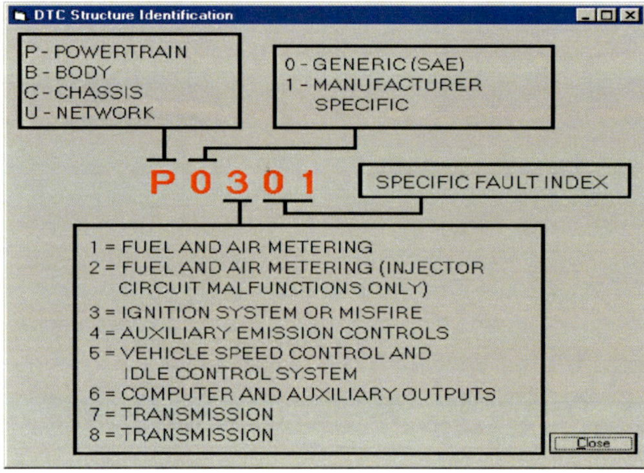

Camshaft Position Sensor

The camshaft sensor sets the cylinder firing sequence and coordinates the fuel injector and coil firing cycles. It works together with the crankshaft position sensor to control engine timing and signals the correct RPM to determine the engine speed

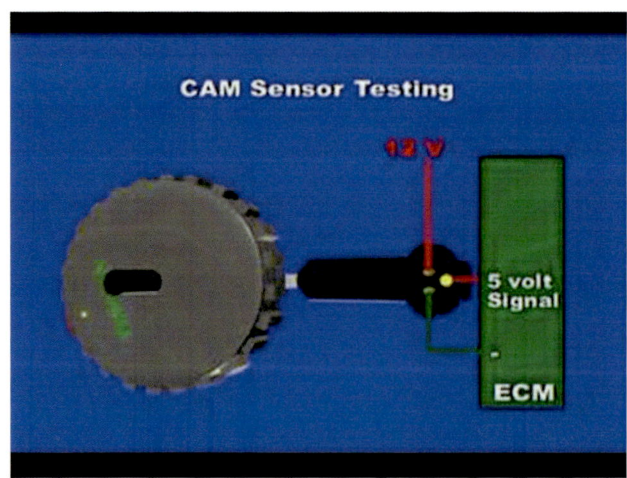

Camshaft Position Sensor CMP

The camshaft sensor sets the cylinder firing sequence and coordinates the fuel injector and coil firing cycles. It works together with the crankshaft position sensor to control engine timing and signals the correct RPM to determine the engine speed

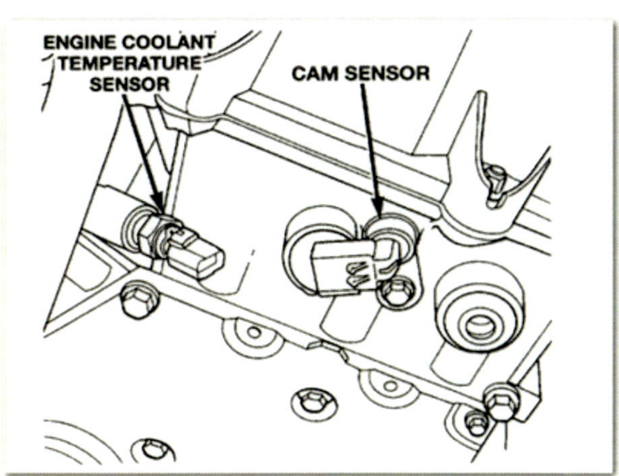

CMP Sensor Location

The camshaft sensor is usually located in the cylinder head of the engine. Part of it is shaped like a cylinder shaft and fits into the head. The crankshaft sensor is typically in the timing cover or on the side of the engine block. It also has a cylindrical section that fits into the block

Symptoms of a Bad Camshaft Sensor

- A noticeable loss of engine power
- Lock the transmission in a single gear until you turn off and restart the engine
- Boat jerking back and forward while losing power
- Irregular acceleration, misfiring, hard starting, or surging
- Prevent ignition spark, so that the engine won't start at all

CMP Sensor Testing

First, check the CMP sensor electrical connector and wires condition. Unplug the connector and check for rust or contamination, like oil, that is interfering with good electrical contact. Then check for wire damage: broken wires, loose wires, and signs of burns caused by nearby hot surfaces. Also, make sure the sensor wires are not touching spark plug wires or ignition coils, which may interfere with the sensor's signal.

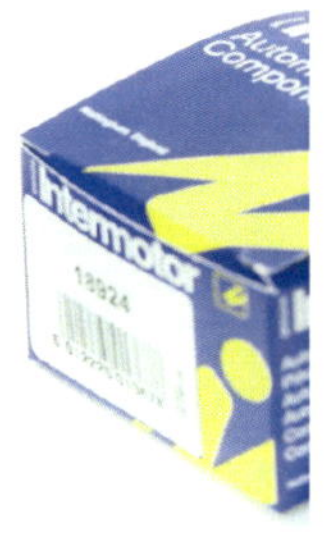

Checking a Two Wire Sensor

- If you have a two-wire, magnetic-type CMP sensor, set your multimeter to "AC volts."
- Check for the presence of power flowing through the circuit. Touch one of your probes to ground (any metal part on the engine) and the other probe to each one of the sensor wires. If neither wire has current, there's a failure in the sensor's circuit
- Have your assistant crank or start the engine

Checking a Two Wire Sensor

- ouch one of your meter probes to either one of the sensor wires and the other probe to the other wire. Check your meter display and compare your reading to your manual specifications. In most cases, you'll see a fluctuating signal between **0.3 volts and 1 volt.**
- If there's no signal, you have a bad CMP sensor.

Checking a Three Wire Sensor

Once you identify the power, ground, and signal wires using your engine repair manual, test the sensor's circuit by setting your multimeter to "DC volts."

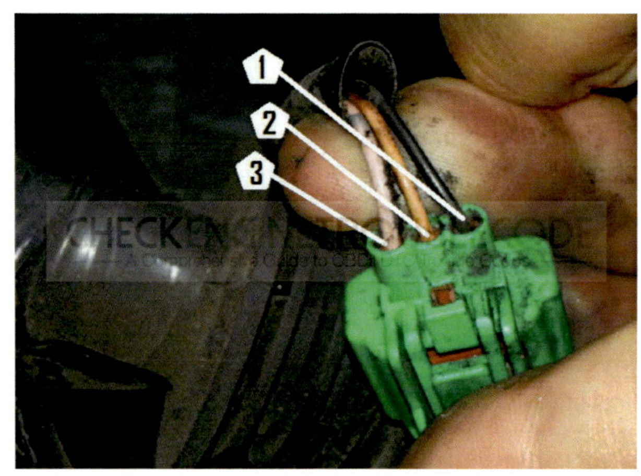

Checking a Three Wire Sensor

- Have an assistant turn the ignition key on, but don't start the engine

- Touch the black probe on your meter to ground (a metal bracket, bolt, or metal surface on the engine itself) and the other probe to the power wire. Compare your reading to the specification in your manual.

- Have your assistant crank or start the engine.

Checking a Three Wire Sensor

Touch the signal wire with the red probe from your meter and the ground wire with the black probe. The reading should be arround 5.0 volts or according with the engine specs, if no signal comes out of the sensor, most likely the sensor is bad.

Symptoms of a Bad CMP Sensor

If you can't find anything wrong with the sensor or its circuit, it's possible you may have an intermittent failure or a failure in a related component. For example, you may have a weakened or overstretched timing belt or timing-belt tensioner or some of the following issues

- Missing during acceleration
- Coil failure
- Detonation
- Short spark plug life
- No spark

CMP Sensor Codes

The OBD II codes related with a failure of an CMP sensor are :

- P0340 to P0344 (Generic Codes)
- DTC Code 41,42 (Mercruiser Scanner Tool)

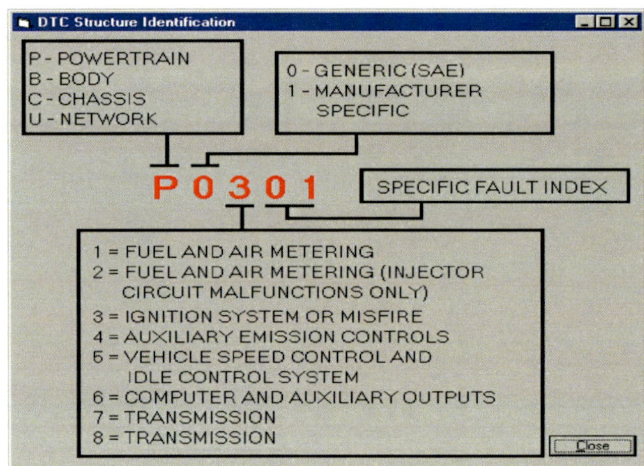

Oxygen Sensor

- An **oxygen sensor** (or *lambda sensor*) is an electronic device that measures the proportion of oxygen (O_2) in the gas or liquid being analyzed

- The sensor is about the size and shape of a spark plug and protrudes into the Exhaust elbow after the risers. It determines if there is a lot or a little oxygen in the exhaust, so the PCM can make adjustments to the amount of fuel being used in the engine to run at maximum efficiency

158

Symptoms of a Bad Oxygen Sensor

- A decrease in fuel efficiency
  - This can happen because of a fuel mixture that is either too lean or too rich. Such a swing in A/F ratio is a sign that an upstream or control sensor is faulty
- A rough idle, A misfire, and/ or hesitation
  - *Note: Keep in mind, however, that these issues can also have other causes that have no relation to the health of an engine oxygen sensors. Therefore, none of them alone is cause enough to replace one*

Common O2 Sensor Failure Causes

- Improperly location into the exhaust elbow
  - This is a typical failure due to constant exposure with salt water. The sensor shall be installed before the exit of salt water
- Age and high mileage/Hours
  - All oxygen sensors are exposed to a constant stream of harmful exhaust gasses, extreme heat and high velocity particulates. Because of this, their efficiency will inevitably decrease over time
- An internal contaminant (poisoning)
  - Fuel mixed with unauthorized additives
  - Dirty oil or contaminated with bad additives (Antifreeze/Silicone)
- An electrical issue
  - A broken sender , A failure at the CPU pin corresponding with that sensor , A low/high voltage coming from the CPU, A bad grounding

O2 Sensor Codes

The OBD II codes related with a failure of an Oxygen sensor are :

- PO171 to PO175 (Generic Codes)
- Code 13 (Mercruiser Scanner Tool)

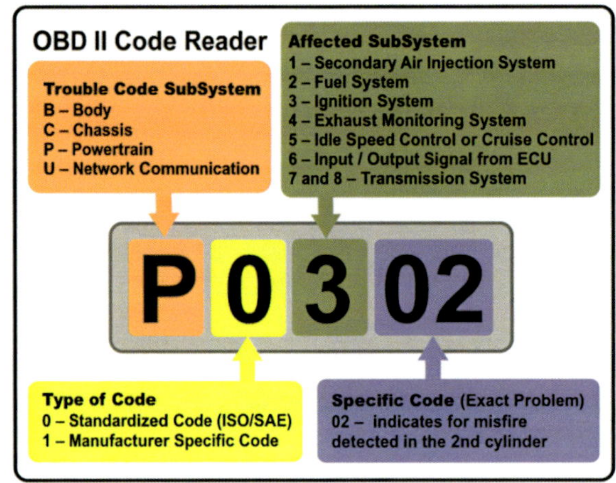

Throttle Position Sensor TPS

The **Throttle Position Sensor** helps ensure your engine is getting the right mixture of fuel and air at every moment. The throttle position sensor provides the most direct signal to the fuel injection system of what power demands are being made of the engine

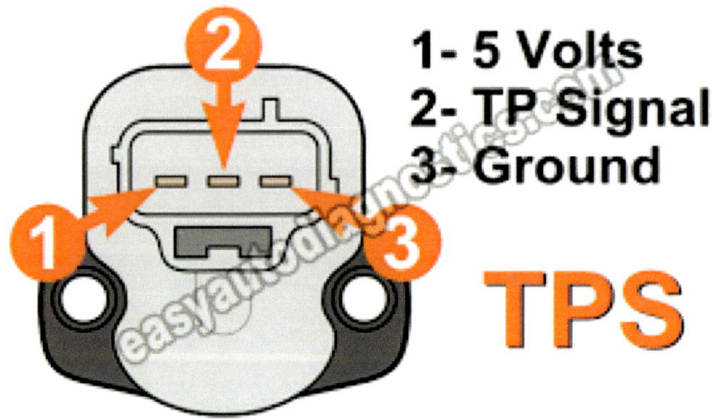

Throttle Position Sensor TPS

The TPS signal is measured and combined many times per second with other data such as air temperature, engine RPM, (actual) air mass flow, and how quickly the throttle position may be changing. These data determine precisely how much fuel to inject into the engine for the next few

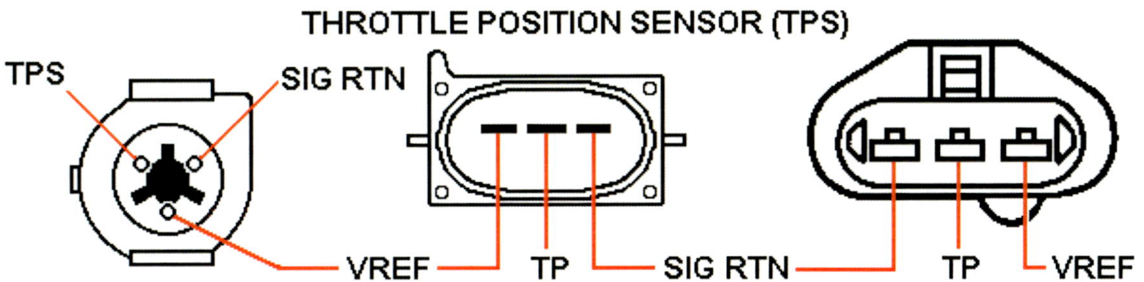

TPS Location

It is located in one end of the butterfly plate axis ,is used to monitor the position of the throttle butterfly valve.

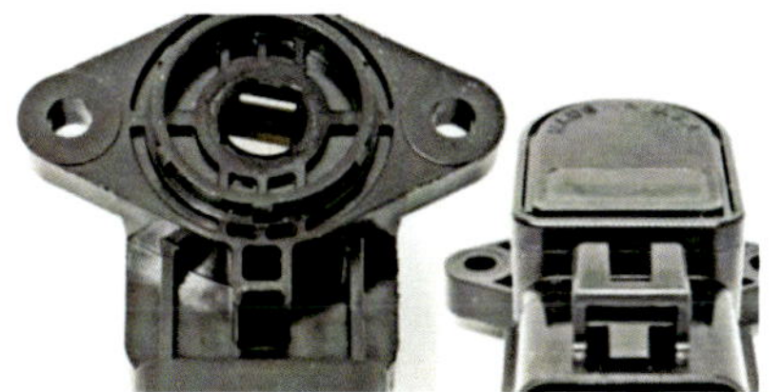

Symptoms of a Bad TPS

- Engine won't accelerate or lacks power when accelerating
  - This fault is related with a defective ingestion of Fuel or Air. Before checking the sensor , verify the fuel and air sources

- Engine won't idle smoothly, idles too slow, or stalls
  - Once again an unbalanced fuel/air ratio affect the idle conditions. Also mechanical problems at the throttle body and poor electrical signals

- Engine accelerates, but won't exceed a relatively low speed
  - Additional to the previous recommendations; this is a typical issue with pressure , temperature and volume of air at the intake manifold

- Check Engine Light comes on, accompanied by any of the above behaviors
  - The TPS sensor works directly with the O2 ,MAP,MAF,and CTS sensors. Before checking those sensors verify electrical connections

TPS Scan Tool Range

- **Voltage**.- Between 0.00 and 5.00 Volts - This is the voltage being monitored by the ECM on the TP sensor signal circuit.

- **Angle**.- Scan Tool Range 0% to 100% - TPS Angle is computed by the ECM from the TP Sensor voltage. TPS Angle should display 0% at idle and 100% at wide open throttle

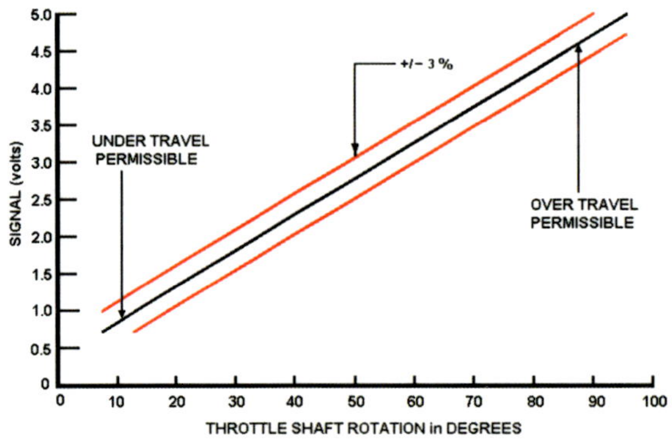

162

TPS Sensor Codes

The OBD II codes related with a failure of a
TPS sensor are :

- PO120 to PO123 (Generic Codes)
- TDC Codes: 21 , 22 (Mercruiser Scanner Tool)

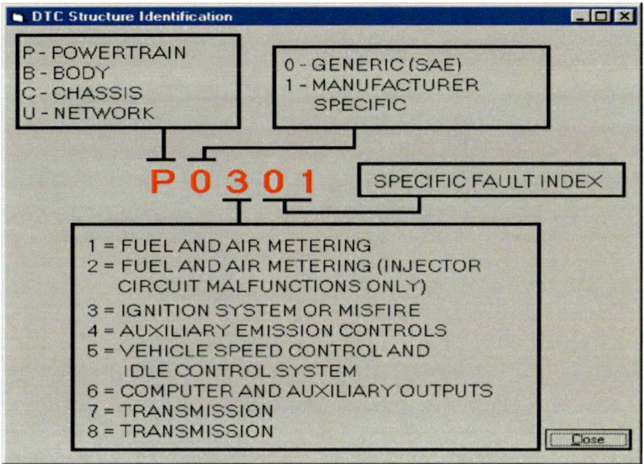

Knock Sensor (KS)

The knock sensor in an engine is used to send
signals to the ECM if this action is detected, thus
preventing a detonation from happening. In a sense,
the knock sensor is a microphone placed in the
engine to listen to any unusual noises that the engine
makes

Knock Sensor (KS)

A **knock sensor** is a transducer used to prevent detonation or spark knock. A spark knock or detonation, as it is commonly referred to, is a form of combustion that can cause different kinds of damage and failure to the engine

Bosch 0261231046

Knock Sensor (KS)

The knock sensor essentially has two uses. The first is to detect any spark knock for an optimum engine performance and the second is to protect the engine from power-robbing

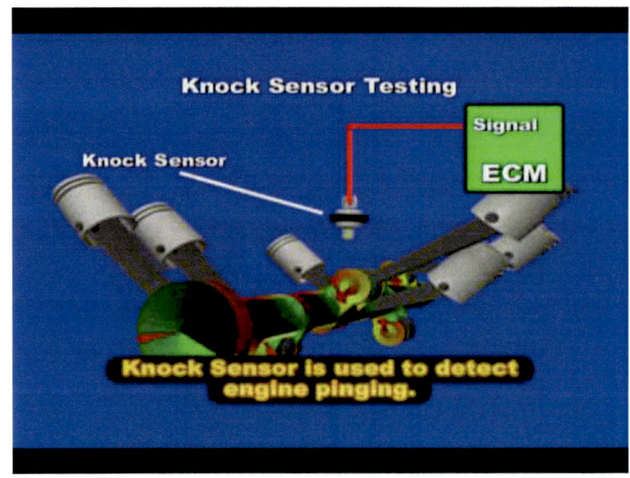

Where is Located KS Sensor

A knock sensor is often located bolted on the center of the wall of the cylinder block. There are also some vehicles where the knock sensor is installed on the cylinder head itself

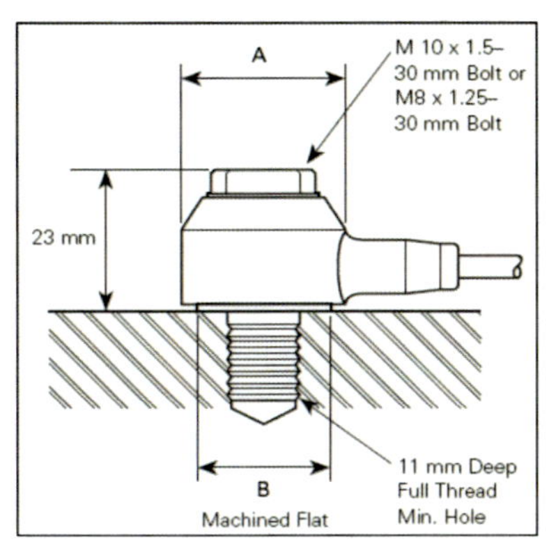

How a Knock Sensor Works

- The knock sensor works by sensing the "knocks" or noises the engine makes when the pressure and heat in the engine is too much. When this happens, the sensor picks up the noise or the vibration being made

- Low volume reverberations from the engine are common which ranges from 6 to 8 kHz

How a Knock Sensor Works

These low reverberations are picked up by the resonating plates inside the knock sensor. This is then relayed to the piezo electric crystalline element of the sensor. Once the element receives the shock wave or signal, it will generate a small amount of voltage which will go to the ECM

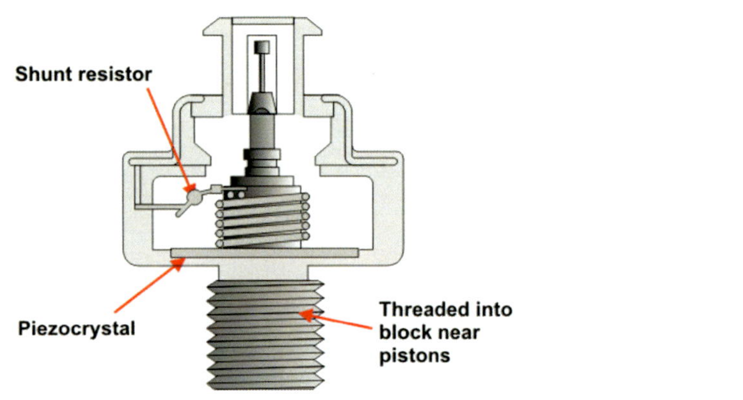

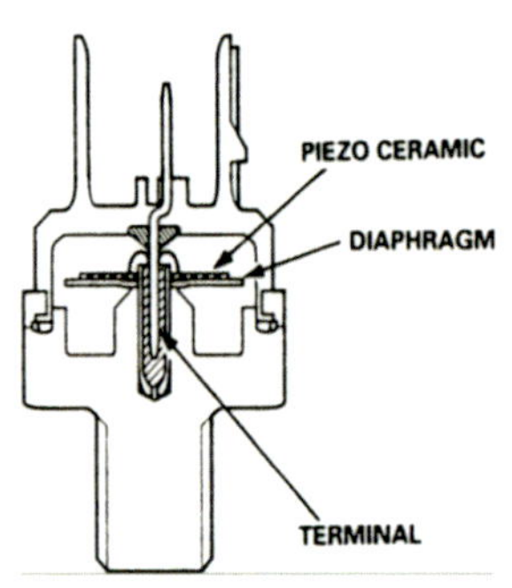

How a Knock Sensor Works

- Knock sensors are used to detect engine pre-detonation (engine knock or ping). The knock sensor (KS) is usually a two wire sensor. A 5 volt reference is supplied to the sensor and there is a signal return from the knock sensor to the ECM

- The sensor signal wire informs the ECM when a knock occurs and in what degree of severity. The ECM will retard the spark timing to avoid pre-detonation. Most ECMs have the ability to learn spark knock trends in the engine during normal operation

How a Knock Sensor Works

- Knock sensors are used to detect engine pre-detonation (engine knock or ping). The knock sensor (KS) is usually a two wire sensor. A 5 volt reference is supplied to the sensor and there is a signal return from the knock sensor to the ECM

- The sensor signal wire informs the ECM when a knock occurs and in what degree of severity. The ECM will retard the spark timing to avoid pre-detonation. Most ECMs have the ability to learn spark knock trends in the engine during normal operation

KS Sensor / Timing

- The knock sensor has a direct influence on the engine's ignition timing. The ECM uses the knock sensor signal to retard ignition timing and thereby reduce pinging. It is common to see a 2 degree or so ignition timing retardation in a step effect

- This means that when the ECM sees a pinging engine it retards timing 2 degrees. If the pinging continues then it adds another 2 degrees of ignition timing retardation. The ECM keeps retarding timing in steps until the pinging stops

KS Sensor Signal

Knock sensors in general are sometimes biased at some particular voltage level, usually 5 volts, although some manufacturers use different bias voltages. This means that the ECM provides a voltage on the signal line

How a fault code is produced

- If the PCM determines that the knock is out of the ordinary or that the noise level is abnormally high, P0326 may set

- If the PCM senses that the knock is severe and cannot be eliminated by retarding the spark timing, P0326 may set. Keep in mind that knock sensors cannot distinguish between a knock from pre-detonation or one from engine problems.

Knock Sensor Testing

- If an engine knock can be heard, fix the source of the mechanical problem first then retest. Make sure the proper octane of fuel was used in the engine. Using lower octane fuel than what is specified can cause a ping or pre-detonation

- Unplug the knock sensor and check for water or corrosion in the connector. If the knock sensor has a seal, verify there is no coolant from the engine block fouling the sensor. Repair as necessary

Knock sensor Testing

Testing the knock sensor is a simple mater. Simply probing on the signal wire with an oscilloscope and tapping on the engine block can generate a signal. Remember that it isn't necessary to tap too hard, since damage can be caused

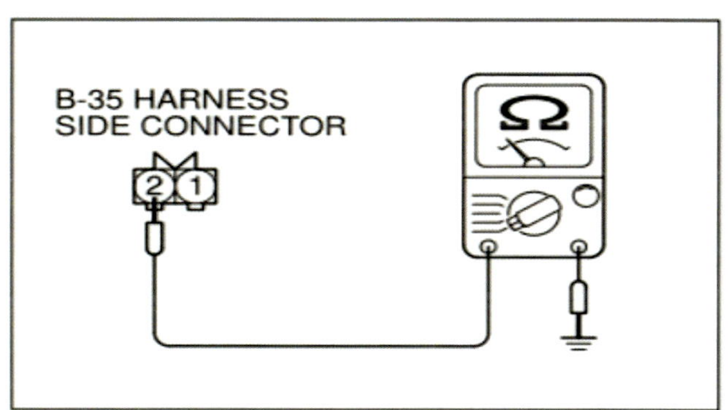

Knock Sensor Testing

- Turn ignition to run position with engine off. Ensure that 5 Volts are present at the KS connector. If there is, check for proper resistance between KS terminal and engine ground (Set Ohm's with engine Off)

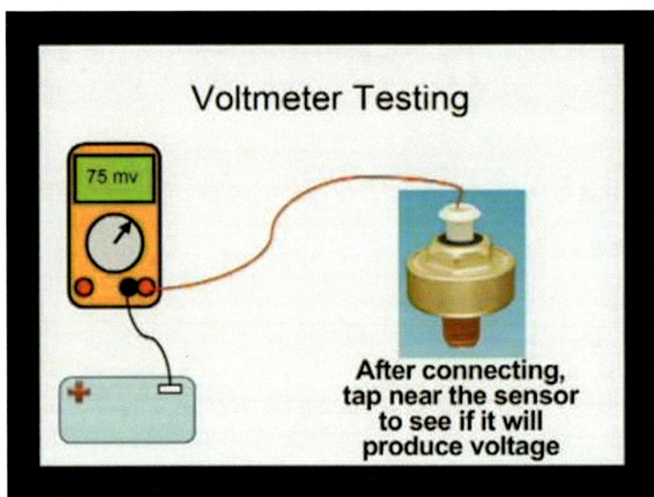

Symptoms of a Bad KS

- MIL (Malfunction Indicator Lamp) illumination
  - Knock sensor connector is damaged
  - Knock sensor circuit is open or shorted to ground
  - Knock sensor circuit is shorted to voltage
  - Knock sensor has failed
  - Moisture in knock sensor connectors
- Audible knocking from the engine compartment
  - Incorrect fuel octane
  - PCM has failed

Symptoms of a Bad KS

- Pinging from engine under acceleration
  - Low Oil Pressure
  - Oil mixed with water
  - A bad Hydraulic lifter
  - Incorrect valve lash calibration
  - A head valves failure
  - Timing chain tensor discharged
  - Timing chain elongated
  - Wear on plain bearings
  - Clutch components failure
  - Transmission bearings wear
  - A broken engine mount
  - Excessive play at the coolant pump bearings
  - Excessive play at the alternator bearings

Knock Sensor Codes

The OBD II codes related with a failure of a TPS sensor are :

- PO325 to PO327 (Generic Codes)
- TDC Codes: 43 , 44 (Mercruiser Scanner Tool)

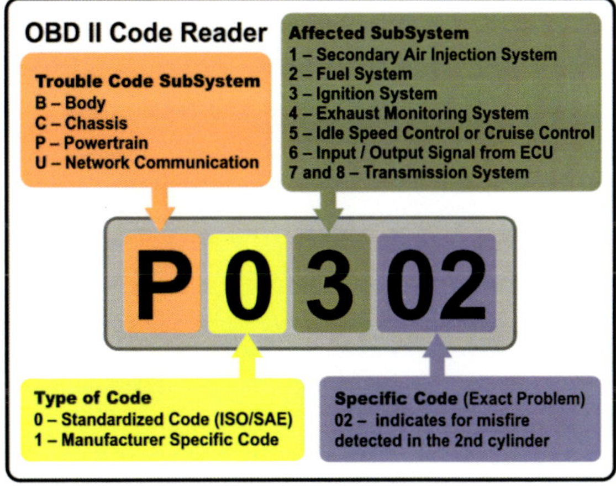

Types of Cooling Systems

- Fresh Water System or Coolant System
  - With heat exchanger that cools the coolant by means of salt water
- Salt Water System
  - The flow of salt water pass through the engine cavities and then out of the engine through the exhaust manifold
- Air Cooled System
  - Closed system with Coolant and Radiator
  - Air Cooled with fins on the head and the Block

Fresh Water System

This is a system where fresh water or antifreeze type coolant circulates inside the engine. This system is similar to a car engine where the antifreeze in the radiator is circulated through the engine

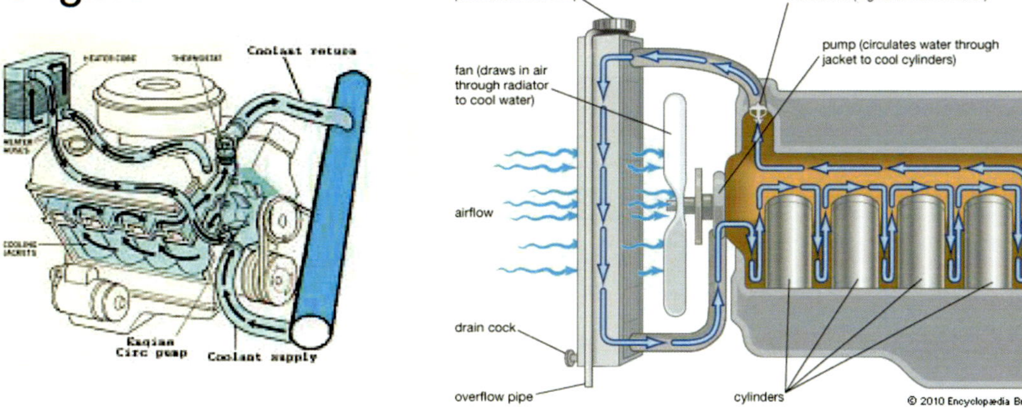

Heat Exchanger

The raw water is brought in an pumped through several small tubes inside the heat exchanger, this tubes will absorb and exchange the heat from the circulating water surrounding them and will exit the heat exchanger towards the exhaust risers to be dumped overboard

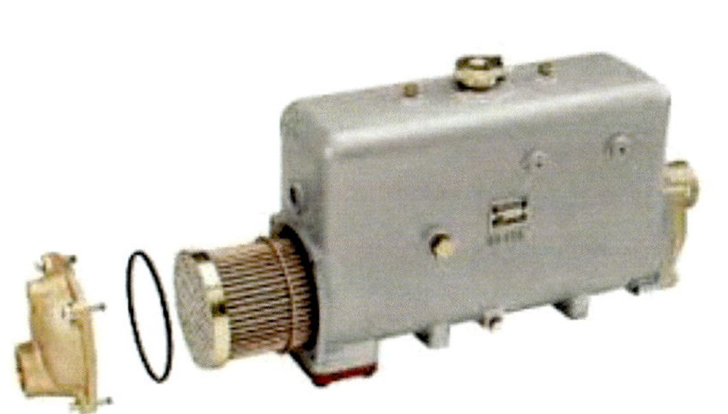

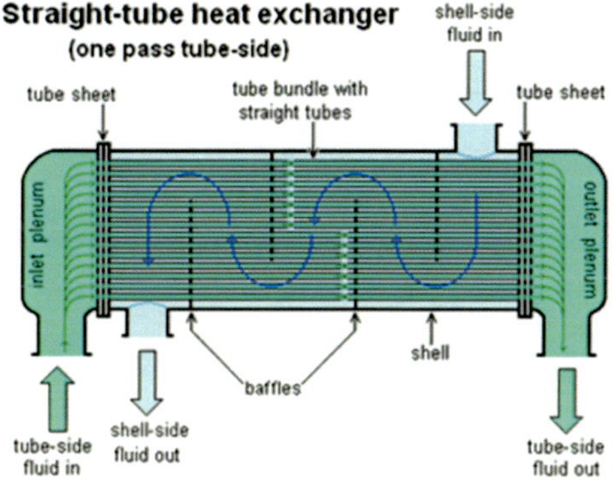

Expansion Tank

Sometimes attached to the heat exchanger itself, sometimes mounted remotely, this tank is a very important part of a closed cooling system. As the engines water coolant gets hot it will expand, increasing in volume. This water needs somewhere to go and the expansion tank is a tank that provides room for this hot water to expanded into

The Half System (Engine Block Only)

The half closed cooling system is designed to circulate cooling water (Coolant) through the block/head only

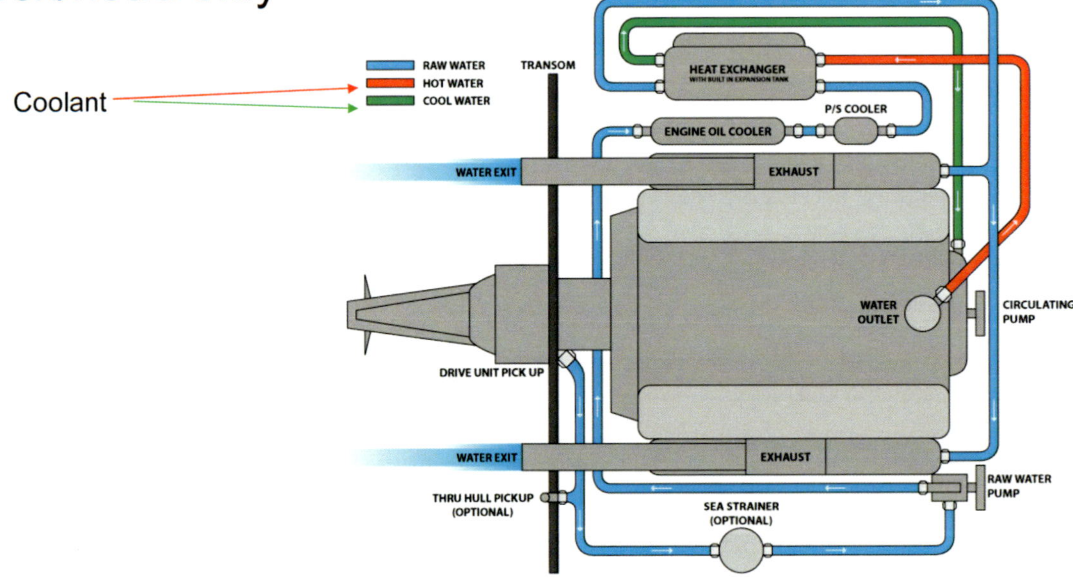

The Half System (Engine Block Only)

The raw water will be used with your Oil Cooler ,Fuel Cooler, Transmission Cooler and Power Steering Coolers as they will not be part of the closed circulating system

The Half System (Engine Block Only)

The raw water then exits the heat exchanger to go through the manifold, into the riser and then exit the boat

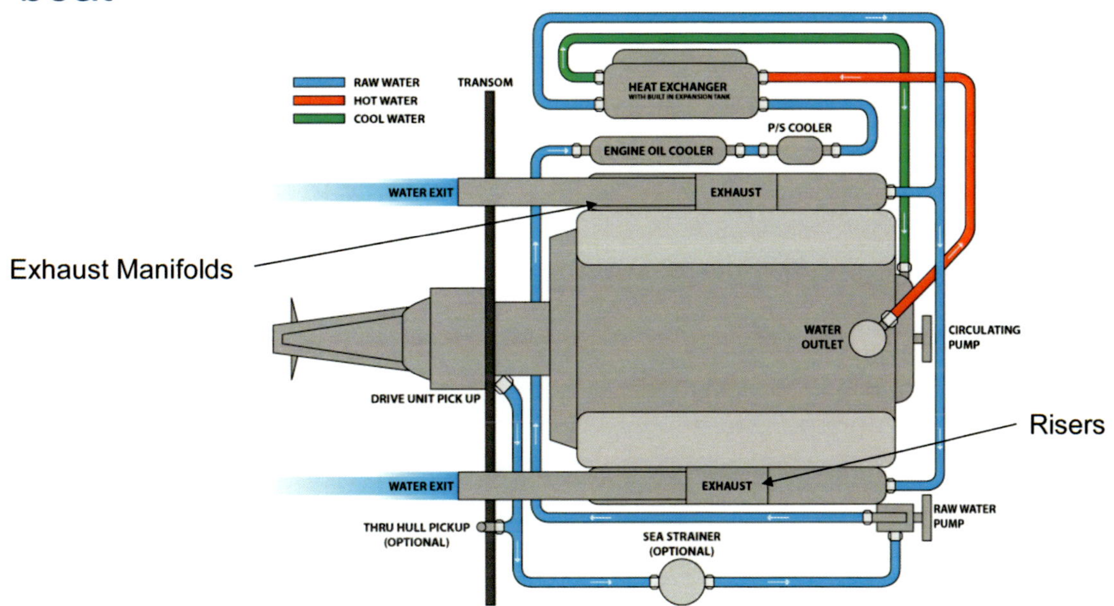

The Half System (Engine Block Only)

As for the circulating coolant in closed cooling system, cool water (Coolant) will enter the engine through the coolant pump and circulate through the engine. Hot water will exit at the water outlet (Therm. Housing) on top of the manifold and head to the heat exchanger to exchange the heat it's carrying with the raw water

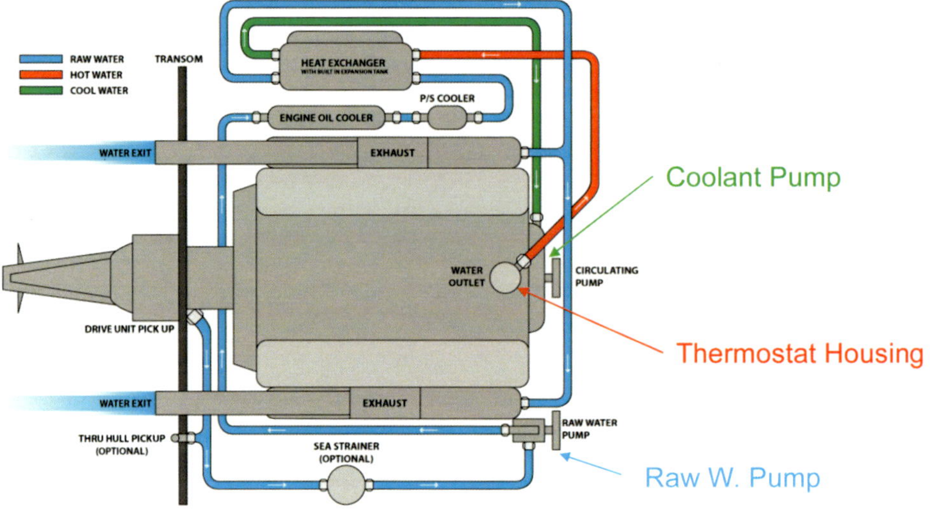

The Full System (Engine Block and Exhaust Manifolds)

The full closed cooling system is designed to circulate cooling water (Coolant) through the block/head and exhaust manifolds

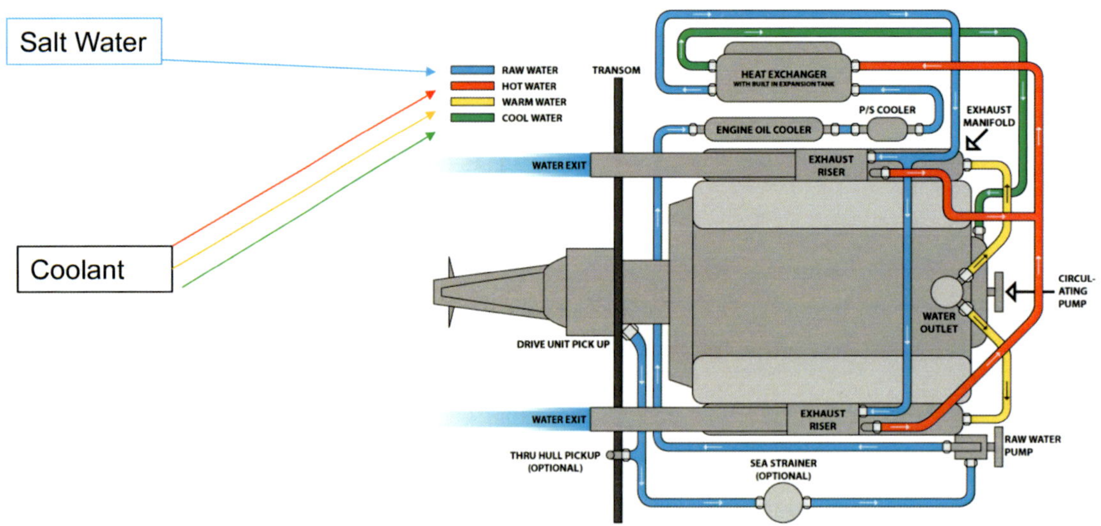

The Full System (Engine Block and Exhaust Manifolds)

With respect to the fresh water (Coolant) cycle, **cool water** will enter the engine through the circulating pump and circulate through the engine

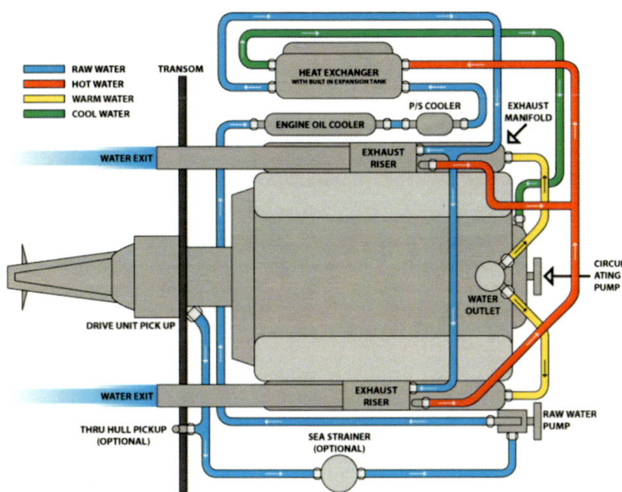

The Full System (Engine Block and Exhaust Manifolds)

The water will heat up in the engine and the **warm water** will exit at the water outlet on top of the manifold and head to the manifolds

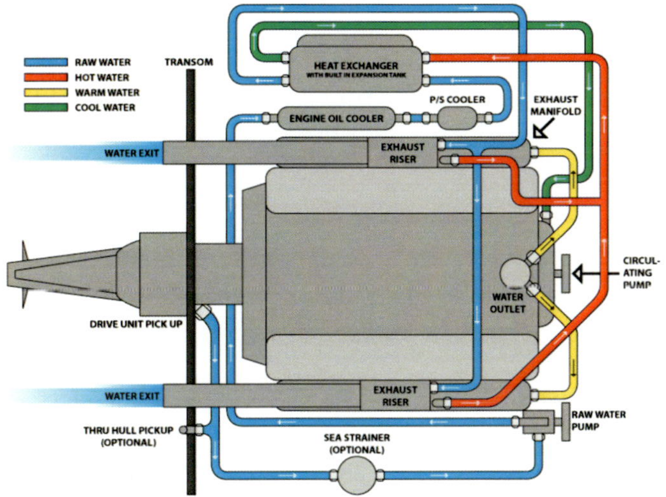

The Full System (Engine Block and Exhaust Manifolds)

After circulating through the manifolds, the **hot water** will head to the heat exchanger to exchange the heat it's carrying with the raw water. After the exchange the **cool circulating water** exits the heat exchanger and returns to circulating pump to begin another loop

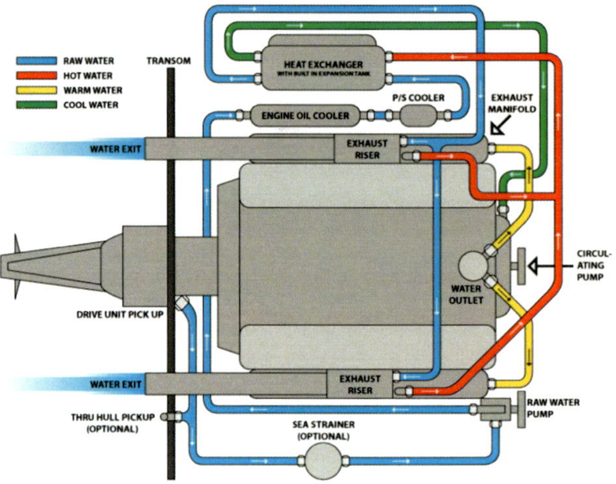

Marine Applications

Application	Half System	Full System
Big Block Chevy to 500 CID	Econo-Kool Freshwater Cooling Kit, Engine Only Big Block Chevy to 500 CID	Econo-Kool Freshwater Cooling Kit, Full System Big Block Chevy to 500 CID
Small Block Chevy to 360 CID	Econo-Kool Freshwater Cooling Kit, Engine Only Small Block Chevy to 360 CID	Econo-Kool Freshwater Cooling Kit, Full System Small Block Chevy to 350 CID
Mercruiser 7.4L BBC	Freshwater Cooling Kit for Mercruiser 7.4 Litre Engines Block Only	Freshwater Cooling Kit for Mercruiser 7.4 Litre Engines Full System Cooli Kit
Mercruiser 4.3 to 5.7L SBC	Freshwater Cooling Kit for Mercruiser 4.3 - 5.7L Carbureted Engines, 1986 thru 1995, Block Only	Freshwater Cooling Kit for Mercruiser 4.3 - 5.7L Carbureted Engines, 1986 thru 1995, Full Cooling System
Mercruiser 4.3 to 5.7L SBC	Freshwater Cooling Kit 4.3 - 5.7L Carbureted Engines with Serpentine Belt, 1997 thru 2001, Block Only	Freshwater Cooling Kit for Mercruiser 4.3 - 5.7L Carbureted Engines with Serpentine Belt, 1997 thru 2001
Mercruiser 3.0L	Freshwater Cooling Kit for Mercruiser - 3.0L, Engine Block Only	
Volvo 3.0L	Freshwater Cooling Kit for Volvo - 3.0L, Engine Block Only	

Full System (4 Cyl In-board)

Closed Cooling System Flow Diagrams

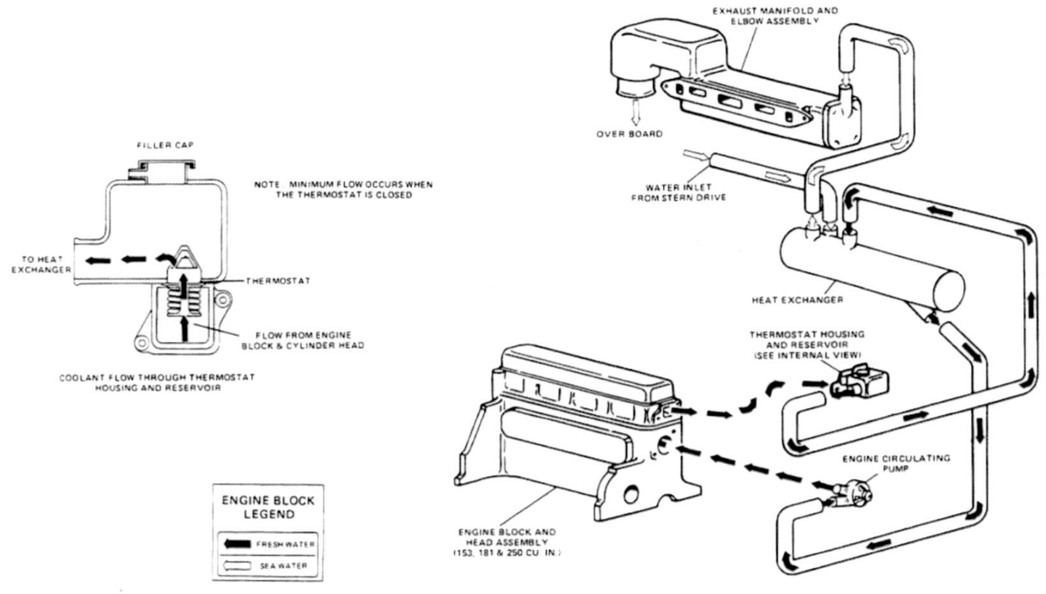

Figure 1. MerCruiser Stern Drive Models 120/140/165 with Closed Cooling

Stern Drive With Outboard Raw Water Pump

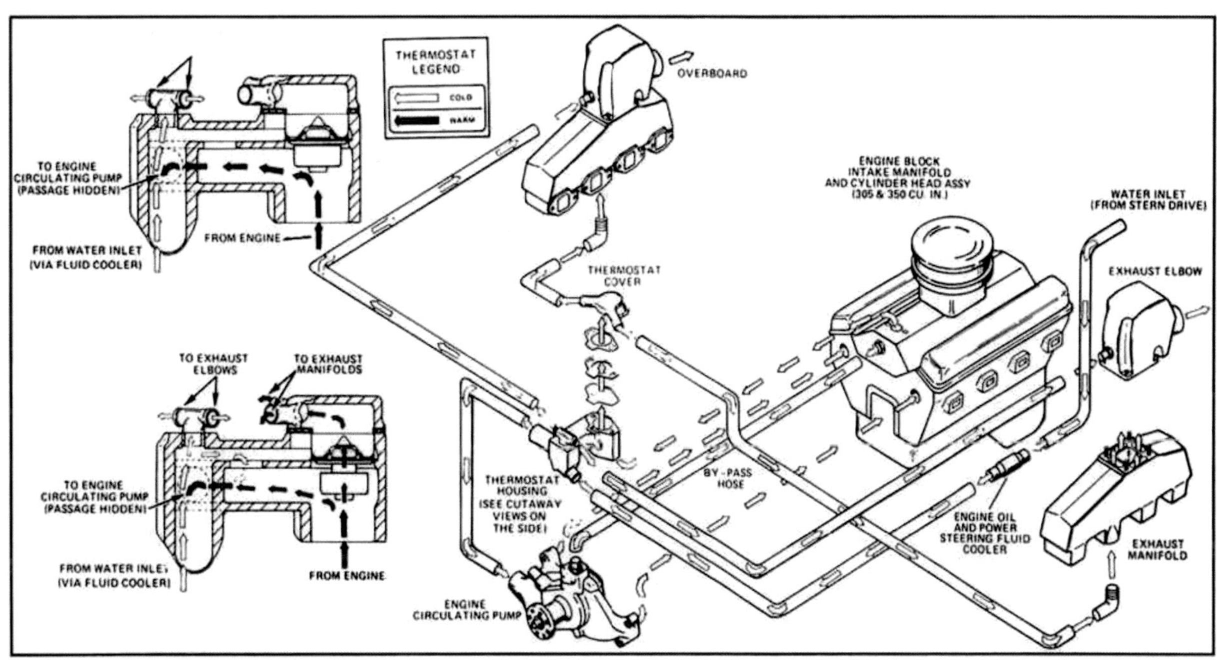

179

Stern Drive With Inboard Raw Water Pump

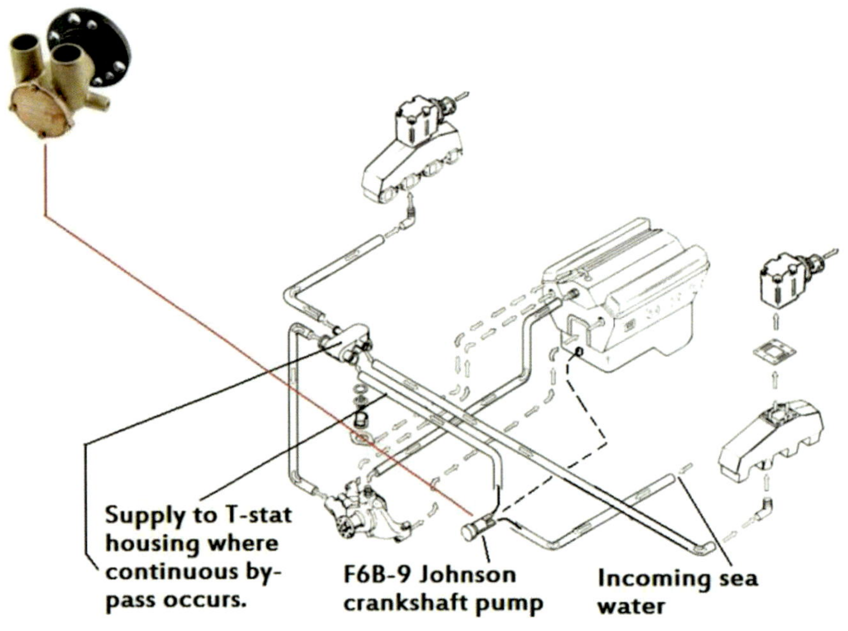

Raw Water System - MerCruiser Stern Drive

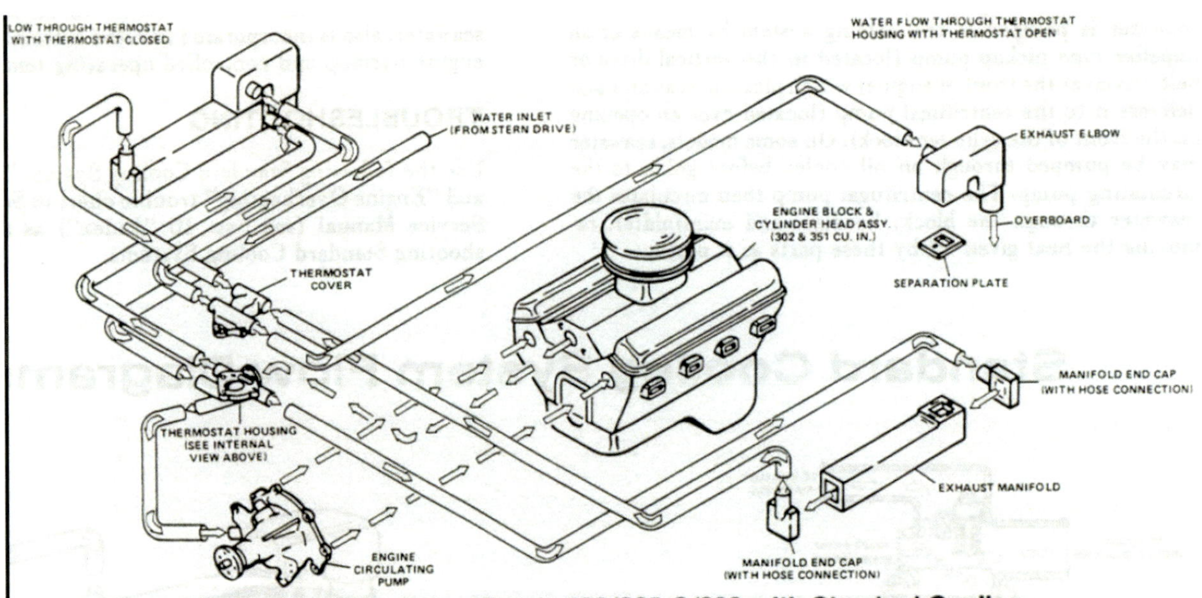

Figure 2. MerCruiser Stern Drive Models 888/225-S/233 with Standard Cooling

Types of Cooling Systems

- The second type of cooling system is called a raw water cooling system

- In a raw water cooling system seawater is circulated through the engine instead of fresh water or antifreeze and that sea water cools the engine directly

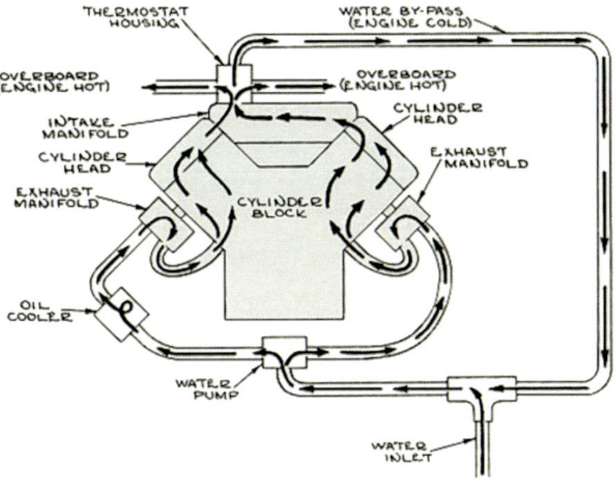

Inboard Raw Water System

In a raw water cooling system seawater is circulated through the engine instead of fresh water or antifreeze and that sea water cools the engine directly. When the boat is used in a fresh water lake this system is excellent but when a boat is used in salt water the salt water circulating inside the engine tends to shorten the life span of the engine

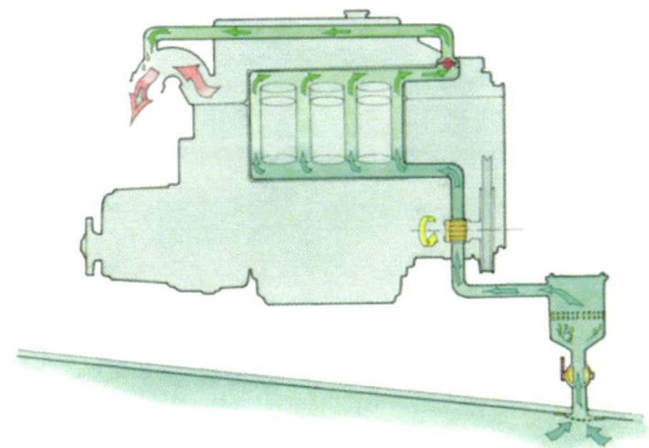

Raw Water Cooling System

- Salt water circulating inside an engine will shorten the life span of the engine.
- When the boat is used in a fresh water lake this system is excellent but when a boat is used in salt water the salt water circulating inside the engine tends to shorten the life span of the engine
- The type of cooling systems becomes a factor in older boats. **Estimated engine life for a fresh water cooled inboard engine is 1500 hours and for a salt water cooled engine is less than 1000 hours**

Stern-Drive Raw Water System

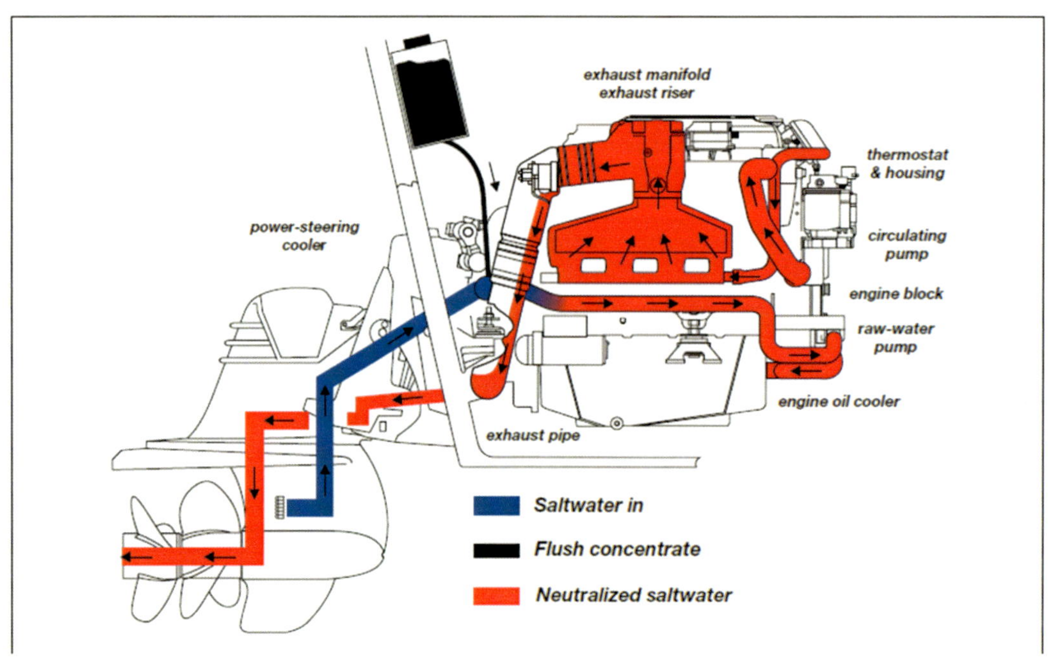

Raw Water Pumps

Salt water Pump and Impeller Kit

Raw Water Pumps

This pump sucks water from the sea. It typically goes through a strainer as it is sucked towards the pump

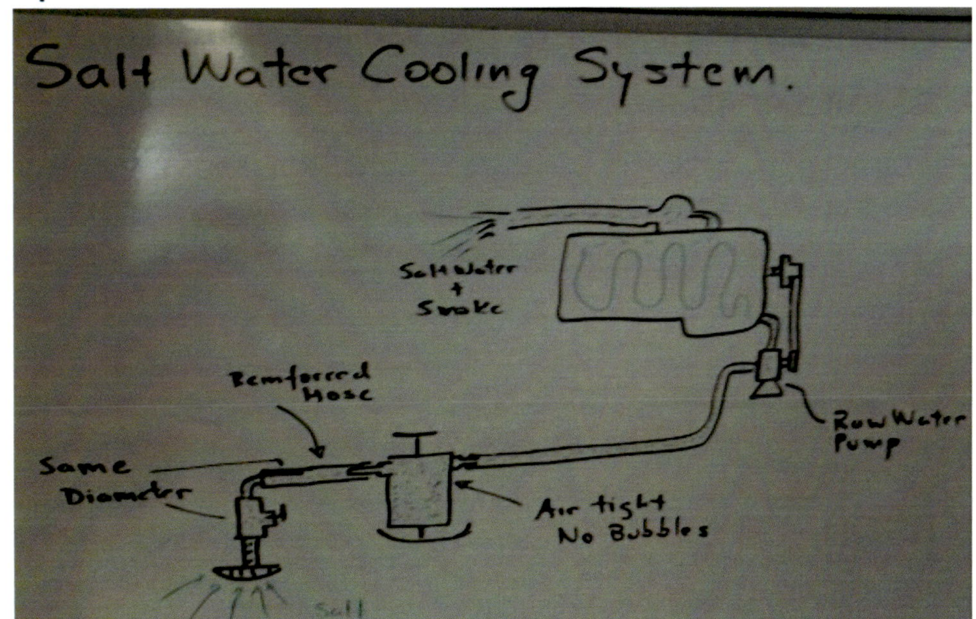

Raw Water Cooling Systems

A flexible impeller pump provides an efficient solution to most raw water pumping needs

Raw Water Cooling Systems

The primary advantage of flexible impeller pumps is that they are self-priming, which means that when the vanes of the impeller are depressed and rebound, they create their own vacuum, drawing fluid into the pump

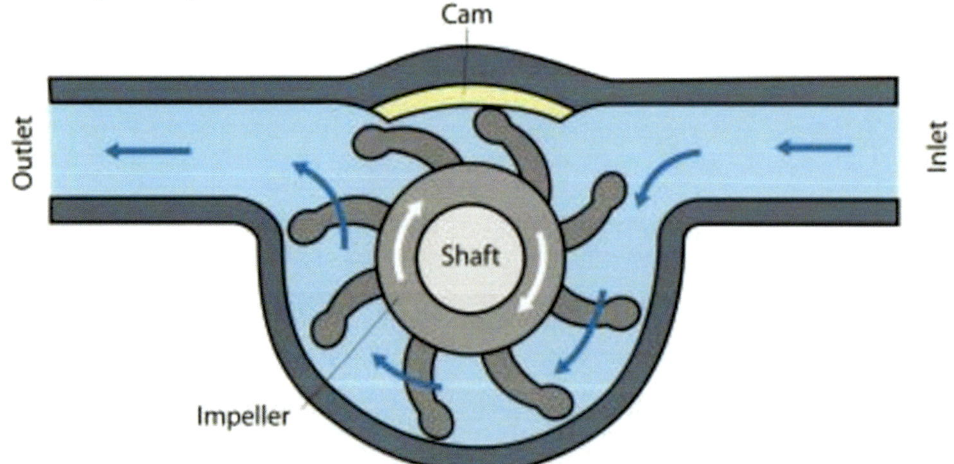

Raw Water pump Housing

The Cam-housing create a reduction in the volume of the fluid to produce a high output pressure

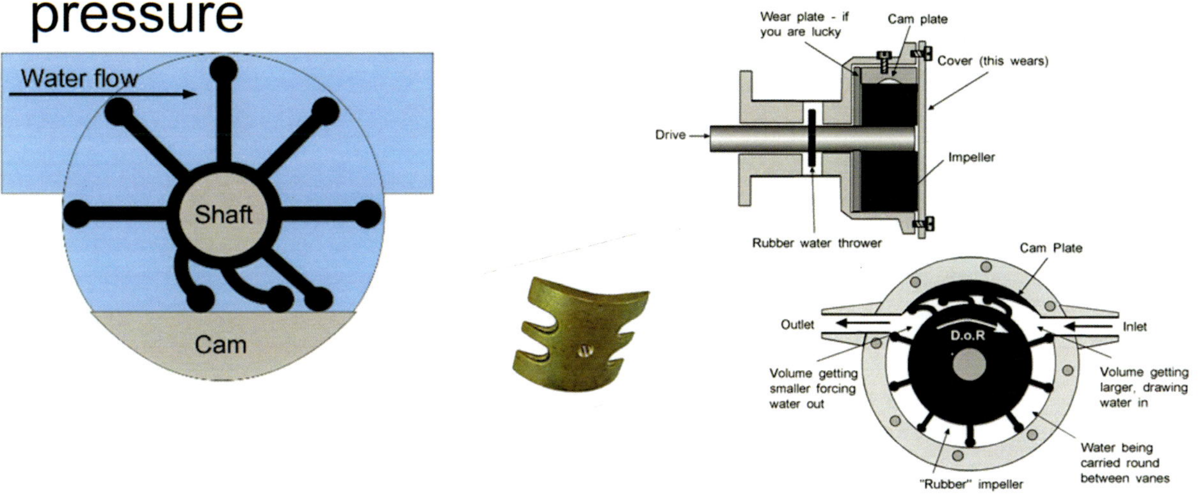

Rubber Impeller / Water line

Thus a flexible impeller pump being used for engine cooling does not need to be manually primed or located below the water line

Impeller Removal Tool

Impeller Removal Tool

The impeller housing can be damage using screwdriver or inappropriate tools

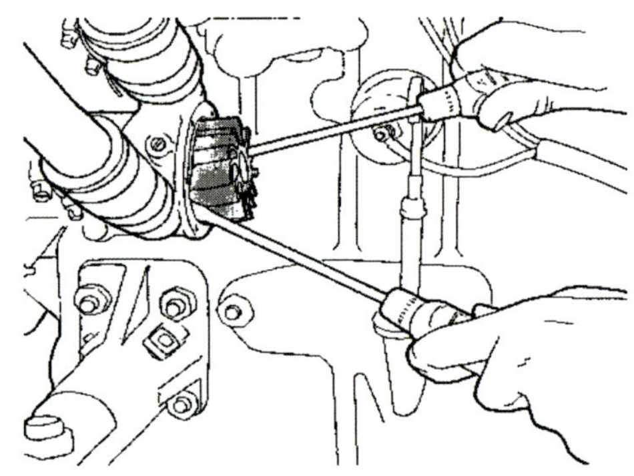

Impeller failure

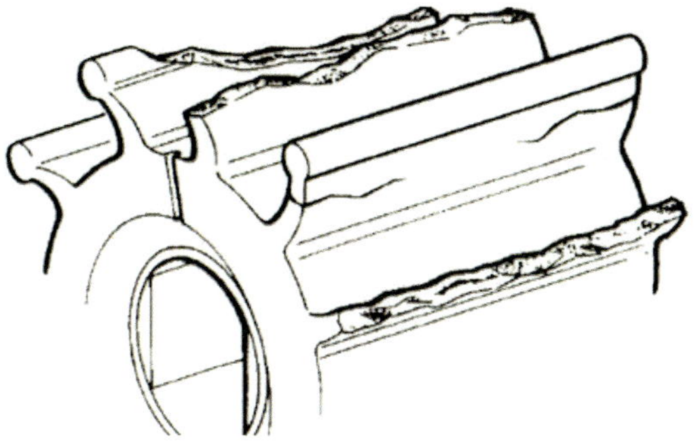

Impeller Plate

Never use screw driver, try to use socket wrench 5/16

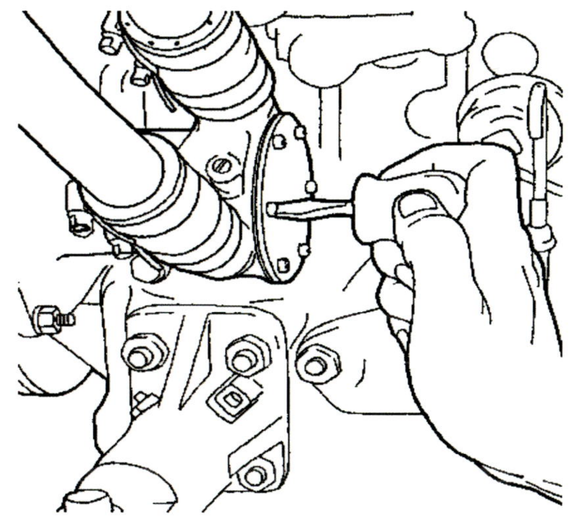

Impeller Plate

Clean the Plate carefully with sand-paper to avoid future leaks

Impeller plate Bolts

Don't try to replace the originals bolts by other similar aftermarket because the thread is different and always will have a leak

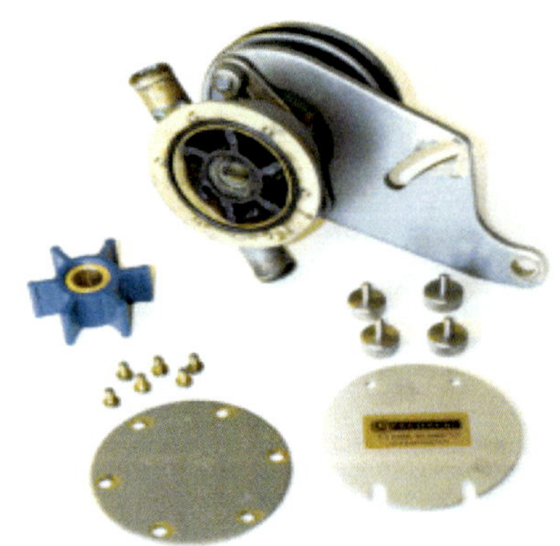 

Seals and O-Rings

Always replace the gaskets and never use silicone, because the excess of silicone could clog the heat exchanger

Impeller Plate installation

Before the plate be tight apply grease in between the center of the plate and the impeller , it is help during the cranking and self-priming period

Debris and Particles

After the impeller replacement clean the input of the heat-exchanger to remove impeller blades fractions and debris

Rubber Impeller / Water line

- An added feature of a flexible impeller pump is that it can pass fairly large solids without clogging or damaging the pump. This reduces the need for filtration of incoming fluids

- For general or salt water applications, a standard long lasting neoprene rubber impeller is used

Impeller Debris

Shaft and Pulley

Remove the belt and verify the shaft by radial play and internal cracking

Raw Water Pump head

A dry pump can lift water up to as much as three meters (9 Ft)

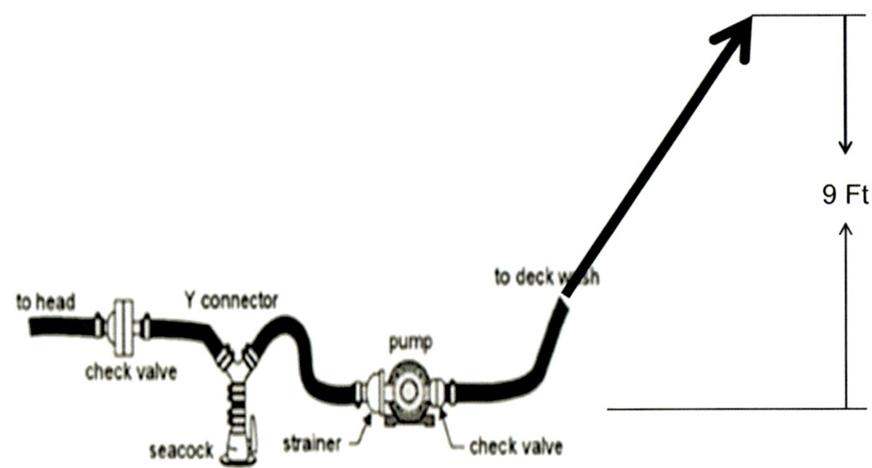

Sea Water Strainer

Traditional strainer were made of bronze with stainless steel strainer baskets and a removable cover to facilitate pulling out the basket for cleaning

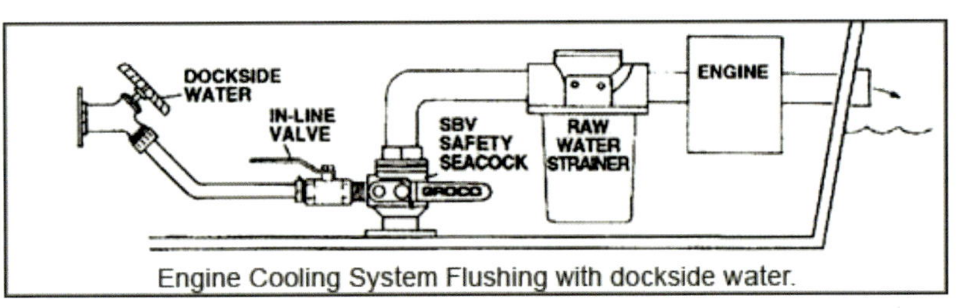

Engine Cooling System Flushing with dockside water.

Sea Water Strainer

The cover have a gasket and it is important that upon reassembly the gasket is seated properly

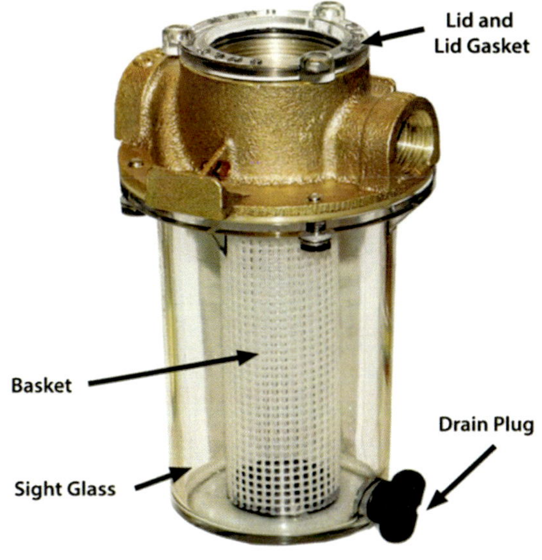

Bronze Strainer

Try to use the same fitting's diameters in both sides to avoid negative suction pressures on the strainer

Raw Water Strainer

Keep the crystal clean to verify easily air bubbles and water contamination

Salt Water & Corrosion

- Running salt water through your engine block and exhaust manifolds will lead to destructive corrosion that is unseen until your engine or exhaust manifolds fail.

- Generally speaking, marine engines cooled with raw water, especially ones that use salt water, have a shorter life span than marine engines cooled with a closed cooling system.

Salt Water & Corrosion

- Same water is pumped through engine block, head, exchangers and returned to ocean either by being dumped into exhaust or by a discharge fitting
- Anodes required in raw water system components sometimes
- Cheapest system
- Engine's lifespan very limited due to deposits and corrosion
  - o Definitely a poor choice - even in fresh water it will kill the engine

Raw Water Cooling Systems

- Raw water strainer and pump impeller are regular maintenance items
  - Raw water strainer should have see-through housing
  - Keep in your spare parts , rubber impellers , raw water pumps , hose clamps
  - When an engine overheats, first thing to check is exhaust water and sweater strainer. After that, impeller
- Always install anti-siphon break in raw water at exhaust elbow

Outboard Raw Water System

Salt water circulating inside an engine will shorten the life span of the engine. In addition overheating problems associated with outboard engines are very common and often result in engine failure

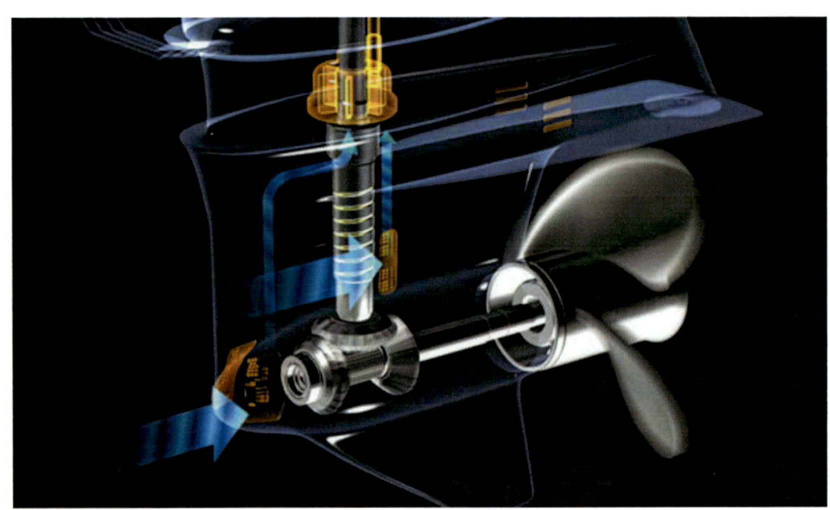

Outboards Raw Water System

Water is inducted through the lower unit by a water pump impeller, and then forced upward to circulate throughout the powerhead, and eventually exits through the exhaust system

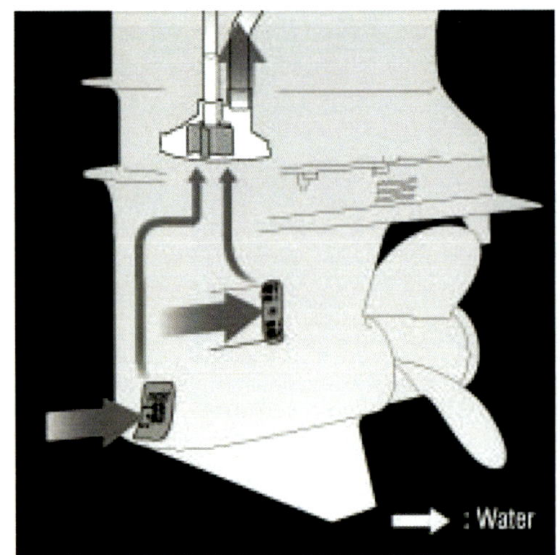

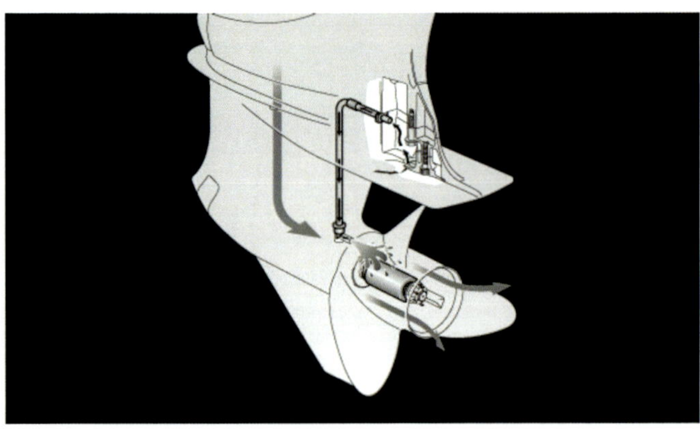

Outboards Raw Water System

Water is drawn into the water pump (A) through the intake opening (B) of the lower unit, where it passes through a cavity that ends up entering the pump itself

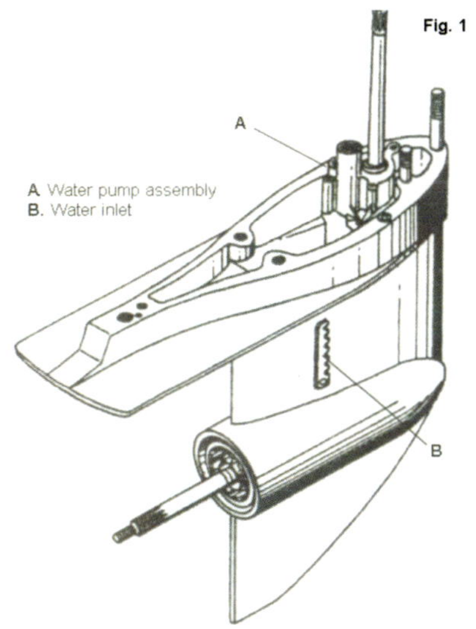

Outboards Raw Water System

The impeller pressures the water into the **long inlet tube** leading to the bottom side of the powerhead. Regularly check water inlet grate to make sure nothing is obstructing water flow

Impeller Pump

The water pump impeller should be replaced every two years, According with the manufacture specifications. An impeller will normally go bad from dry rot attributed to non-usage more so than wearing out from excessive usage

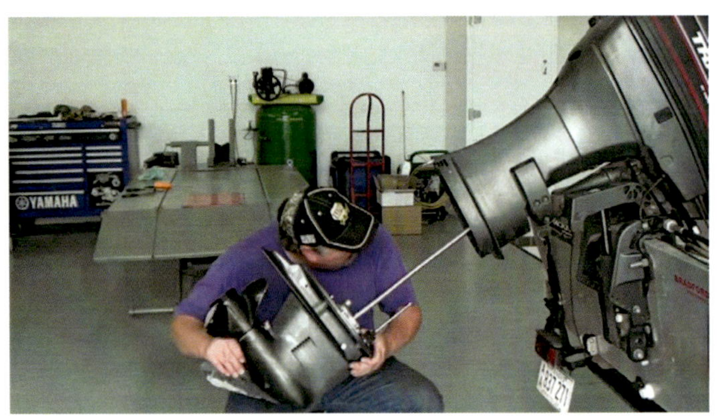

Impeller Warning Horn

One thing to remember on an outboard is the warning horn. You will hear a loud constant shrill whistle if your outboard is in an overheat condition. DO NOT ignore this warning as an outboard can totally seize up within a matter of seconds if the overheat is severe enough

Powerhead Thermostat

Upon reaching the powerhead, water is directed towards the thermostat and <u>bypass valve</u> .

Powerhead Thermostat

The bypass valve acts as a pressure relief until such time the thermostat opens and allows circulation through the powerhead to begin. Most outboard thermostats are factory designed to open at approx. 160 degrees

Impeller Pump & Thermostat

The impeller will be the cause of 95% of outboard overheat situations. It is rare that a thermostat goes bad, but it does occasionally happen

Outboard Thermostat Function

- If an outboard is operated without a thermostat installed, it will surely suffer in some way from poor performance all the way to causing internal damage

- If the engine not obtain the operational temperature, it will run very sluggishly, and also cause fuel/oil mixture to not totally burn, which causes excessive carbon buildup. Carbon buildup in an outboard is one of its worst enemies

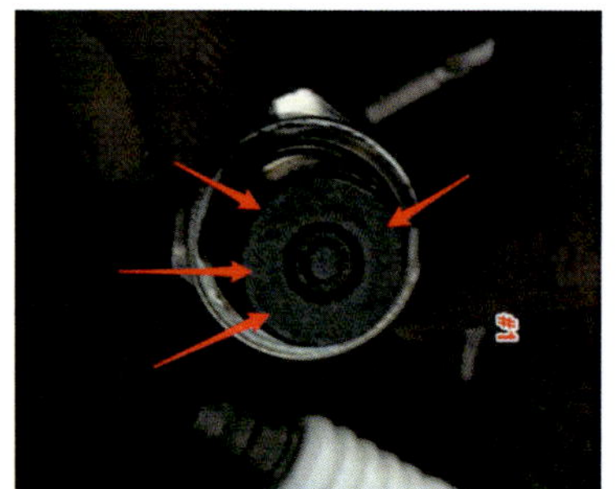

Carbon Deposit Consequences

- This condition usually most often causes problems with the piston rings becoming stuck in the piston ring grooves.

- Once a piston/s is scored, it is only a matter of time before the engine will self destruct unless it is caught in time to perform an overhaul

Flushing Cooling System

The outboard should be flushed after every use in salt or dirty water. Freshwater flushing with clean water will dislodge most contaminants (sand, silt, mud, etc.) and force them out of the passages in the cooling system

Flushing Methods

There are three methods to flush an outboard: using a flush bag, flush muffs or the freshwater flushing attachment built into the outboard

  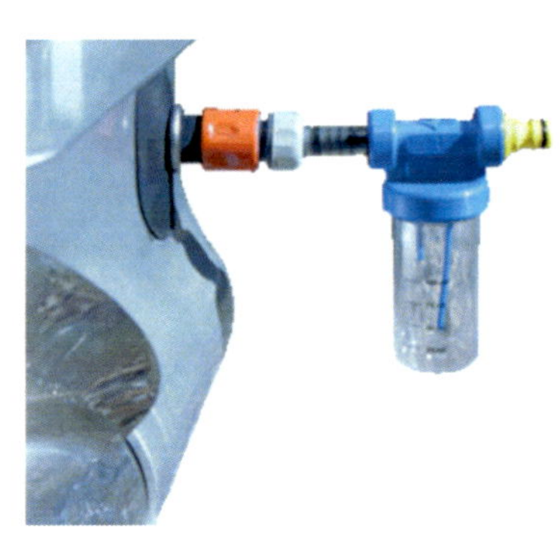

Flush Bag Method

A flush bag is a sturdy collapsible bag that fits under the lower unit of the outboard and fills with water. Submerging the gearcase allows the motor to take in cooling water from the inlet screens and return it to the bag through the propeller

Flush Bag Method

Use the Flush Bag method with the boat on a trailer or the boat moored. It requires running the engine (but only after the bag is filled with water). Outboards with more than one inlet on each side of the lower unit should be flushed using this method

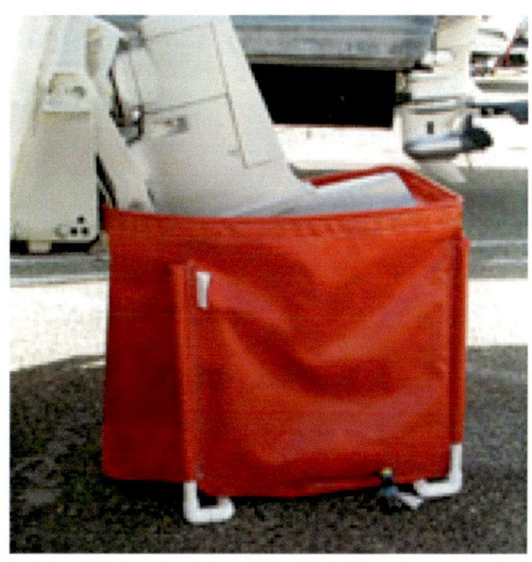

Flush Bag Procedure Steps

- Trim the outboard to the vertical position
- Remove the propeller
- Screw the garden hose into the connector on the bag and fill the bag
- Start the outboard and set the throttle to no more than a fast idle (800-900 RPM max)
- Watch for water coming out of the pilot tube at the lower rear of the cowling. Run the engine for 10-15 minutes
- Stop the engine and turn off the faucet
- Add water to the flush bag while the motor is running to maintain the maximum water level

Flush Muff Method

For models with single inlet lower units

Flush muffs are rubber cups that fit over the water inlet on the gearcase (one of the cups has a hose fitting). They require running the engine (but only after the garden hose is in place and water running).

Flush Muff Procedure Steps

- Trim the outboard to the vertical position and remove propeller
- Screw the garden hose onto the flush muff connector
- Place the flush muffs over the water inlets on gearcase
- Turn the water on slowly, until you can see a bit of water leaking from around the flush muff rubber cups
- Check to ensure the rubber cups are still in place over the inlets

Flush Muff Procedure Steps

- Start the outboard and set the throttle to no more than a fast idle (800-900 RPM max)

- Watch for water coming out of the pilot tube at the lower rear of the cowling. Increase the water flow, if necessary, to maintain a flow of water   from the pilot tube. Note: it may take 20–30 seconds to fill the water passages before water will exit the pilot tube

Built-In Flushing Device

A built-in flushing device is a hose fitting, usually located near the lower rear cowling. Do not run the engine during this procedure

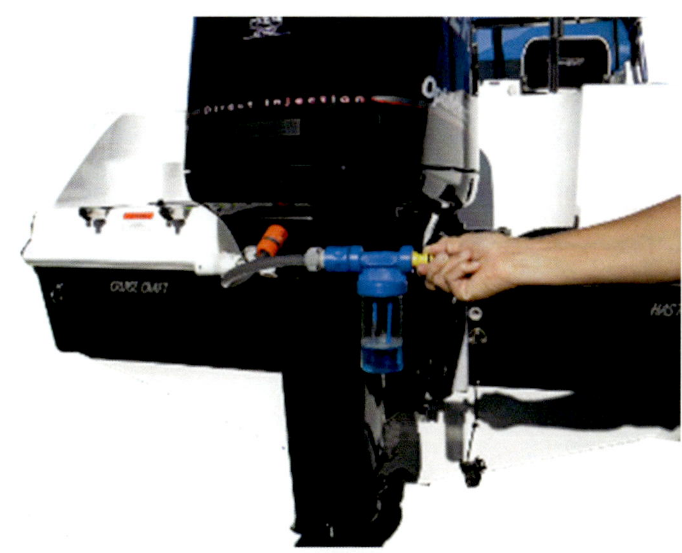

Adding Descaling Removers

Some removers and acids can be added in a separate tank. Consult the service manual in order to verify which products are recommended in each case

Built-In Flushing Procedure

- Connect a garden hose to the outboard's flushing device fitting
- Turn on the faucet; let freshwater run through the outboard for 10-15 minutes
- Turn off the faucet; disconnect
- Put the cap back on engine's flush fitting (per owner's manual)

Flushing Inboard Engines

Many boaters who have inboard engines with fresh water closed cooling systems, do not realize that the raw water side of the engine, where salt water passes through the heat exchanger and out the exhaust manifold, can have major salt corrosion problems and expenses

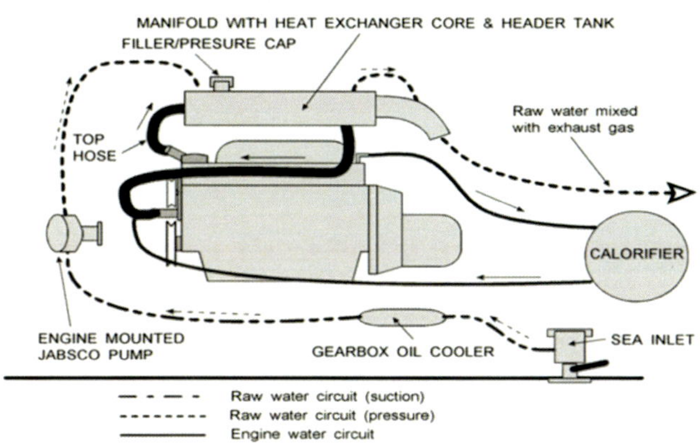

Flushing Inboard Engines

You can use any type of descaling product. According with my experience the following is the most simple way to do the flushing

- 1.- Close the seacock valve and remove the cap from the sea strainer.

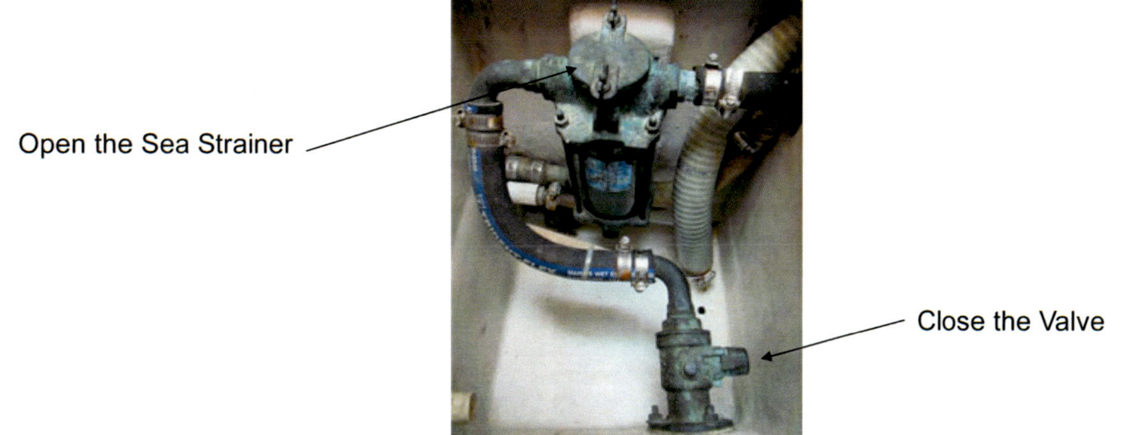

Flushing Inboard Engines

- 2.-Introduce into the opened strainer a hose garden that can be controlled with a ball valve
- 3.- Use the hose garden valve to keep the strainer full of water , then start the engine

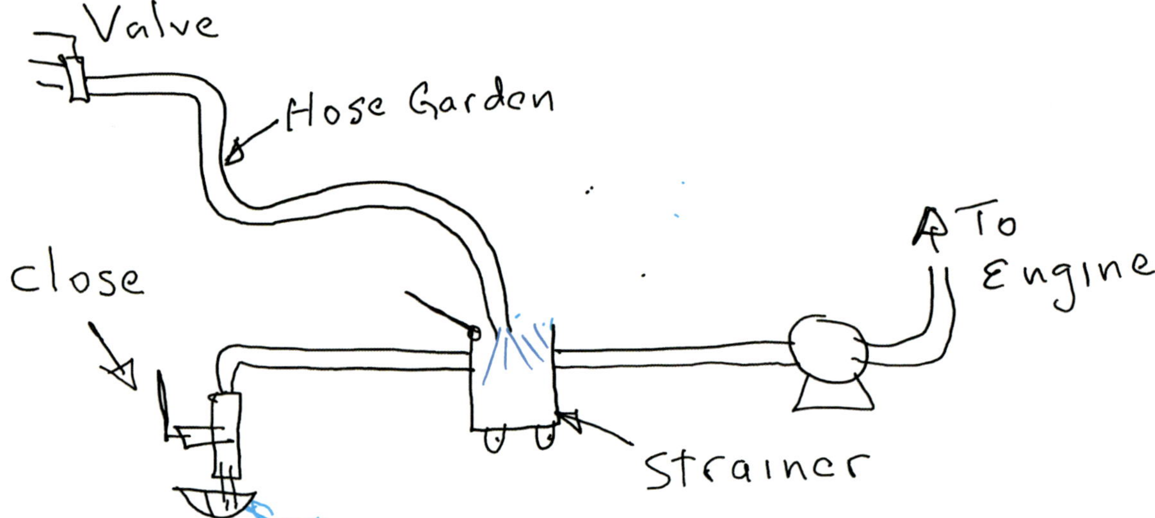

Flushing Inboard Engines

4.- With one hand control the amount of freshwater and with the other one add the flushing product

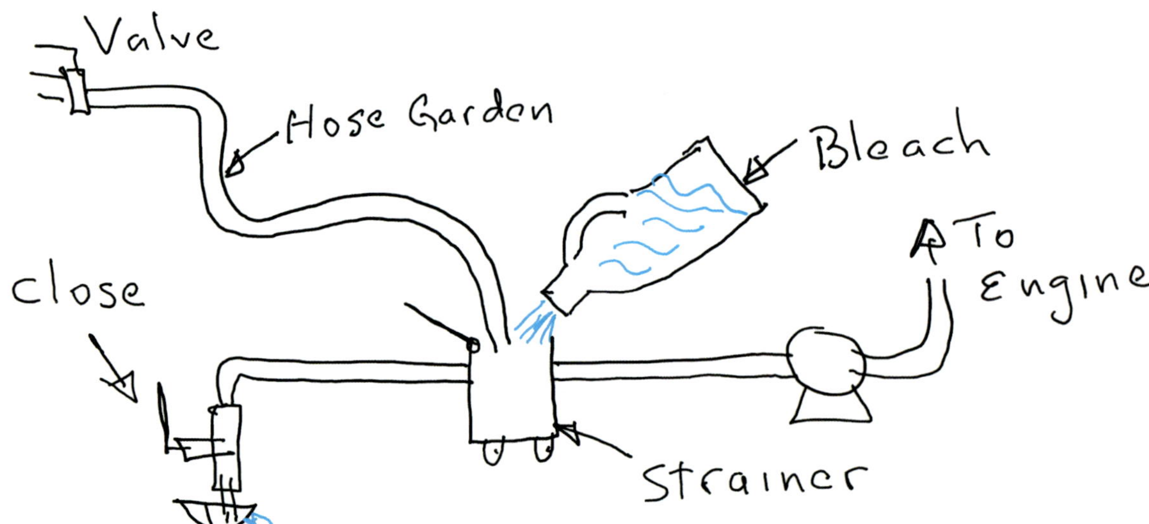

Flushing Inboard Engines

5.- Keep the engine running (idle) with mixture of fresh water and flushing product around five minutes

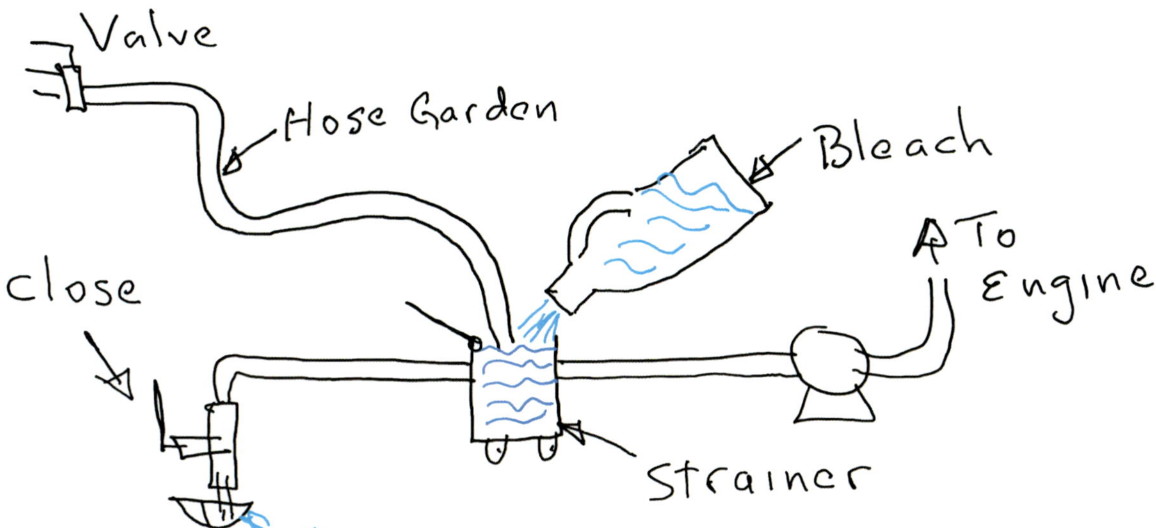

Flushing Inboard Engines

5.- Stop the engine; remove the hose garden and close the sea strainer, then open the seacock valve and re-start the engine

6.- Observe the flow of water at the sea strainer and check the engine temperature

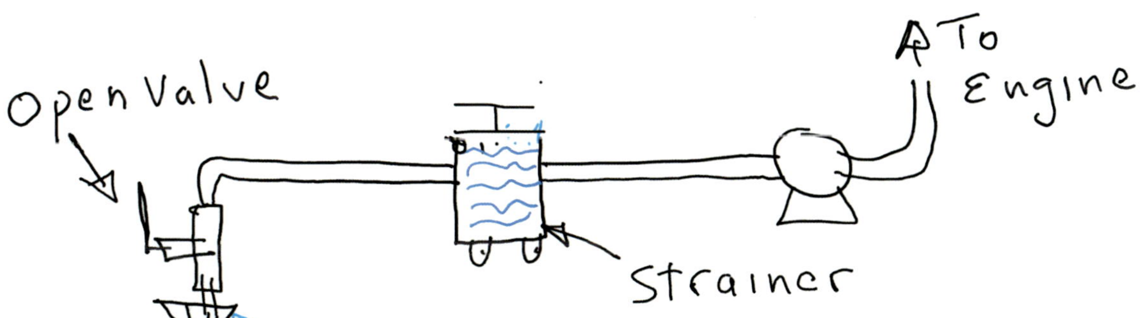

Chapter 6
Lubrication System

Lubrication System
(Inboard Four Stroke Engines)

- ## The Gear Type Oil Pump
  - It consisted of two meshing gears, which rotate inside a close-fitted housing. As the gears rotated they carried oil around, against the housing. The meshing of the gear teeth forced the oil out into the pump

- ## The Rotor Oil Pump
  - This consisted of a rotor, with four or five external lobes, which rotated inside an outer ring (termed a stator) having five or six internal lobes. The axis of the inner rotor was offset from the axis of the outer ring
  Eccentric Rotor Oil Pump

Inboard Gas Engine oil Pump

To avoid the need for <u>priming</u>, the pump is always mounted low-down, either submerged or around the level of the oil in the sump

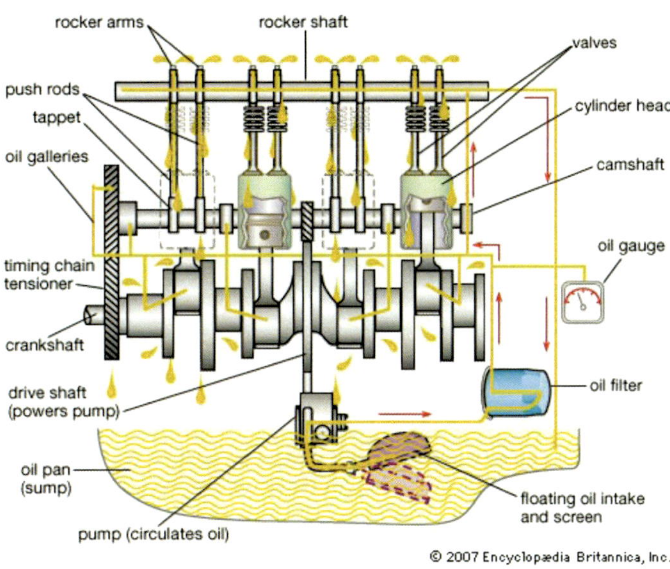

Inboard Engine oil Pump

Placing the oil pump low-down uses a near-vertical drive shaft, driven by helical <u>skew gears</u> from the camshaft

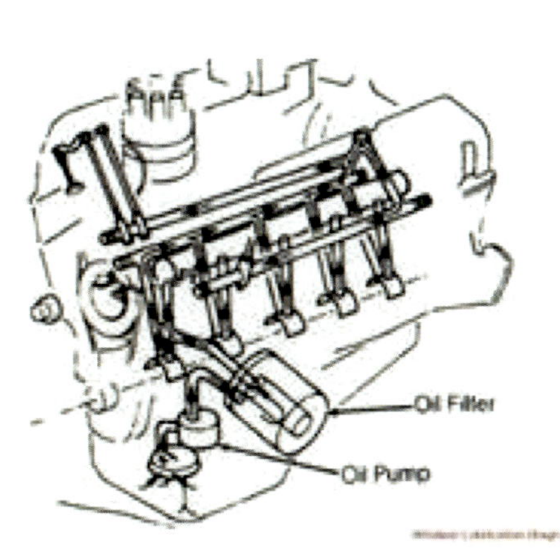

Engine Oil Circulation

- The **oil pump** in a Gas engine circulate the engine oil under pressure to the rotating bearings, the sliding pistons and the camshaft of the engine

- This lubricates the bearings, allows the use of higher-capacity fluid bearings and also assists in cooling the engine

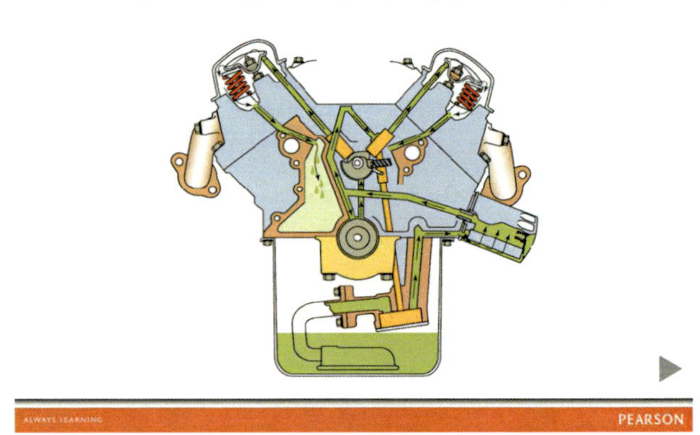

Maintaining Oil Pressure

- The oil pressure at the pump outlet, which is what opens the pressure relief valve, is simply the resistance to flow caused by the bearing clearances and restrictions

- All engines are "open systems", because the oil returns to the pan by a series of controlled leaks

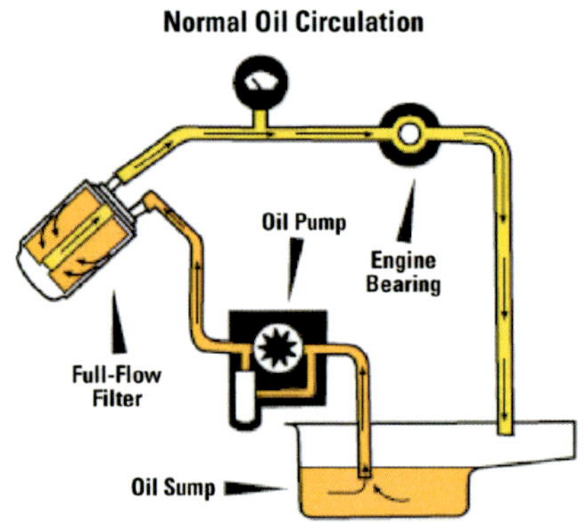

Oil Pressure

- The pressure is actually created by the resistance to the flow of the oil around the engine. So, the pressure of the oil may vary during operation, with temperature, engine speed, and wear on the engine

- Colder oil temperature can cause higher pressure, as the oil is thicker, while higher engine speeds cause the pump to run faster and push more oil through the engine

- Excess bearing clearance increases the pressure loss between the first and last bearing in a series.

Oil Pressure Sensor & Gauge

- Because of variances in temperature and normal higher engine speed upon cold engine start up, it's normal to see higher oil pressure upon engine start up than at normal operating temperatures, where normal oil pressure usually falls between 30 and 45 psi.

- The oil pressure is monitored by an oil pressure sending unit, usually mounted to the block of the engine

Oil Pressure Sensor & Gauge

- Problems with the oil pressure sending unit or the connections between it and the driver's display can cause abnormal oil pressure readings when oil pressure is perfectly acceptable.

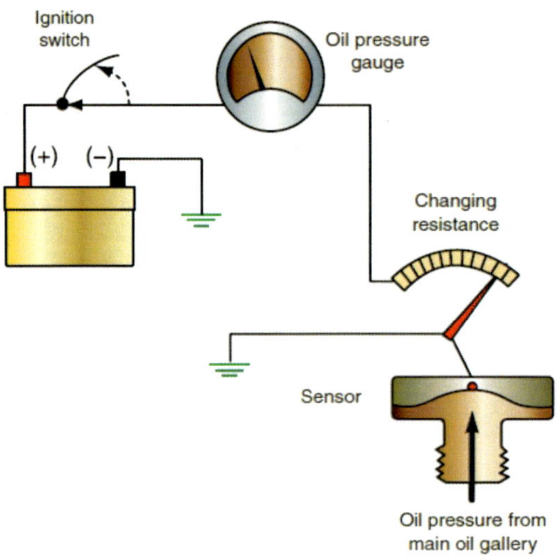

High & Low Oil Pressure

- Too much oil pressure can create unnecessary work for the engine and even add air into the system

- Low oil pressure, however, can cause engine damage. Low oil pressure can be caused by many things, such as a faulty oil pump, a clogged oil pickup screen, excessive wear on high mileage engines, or simply low oil volume

- Indications of low oil pressure may be that the warning light is on, a low pressure reading on the gauge, or clattering/clinking noises from the engine.

- The leading cause of low oil pressure in an engine is wear on the engine's vital parts

Tensioner and Lifters

One of the first notable uses in this way was for hydraulic tappets in camshaft and valve actuation. Increasingly common recent uses may include the tensioner for a timing belt/chain or variators for variable valve timing systems

Timing Chain Tensor

Oil Pressure Sensor & Switch

- Both of them are located at the oil filter bracket. The Oil Pressure sensor is connected with the gauge and the Oil pressure switch is connected in series with the power wire of the ignition system and Fuel Pump

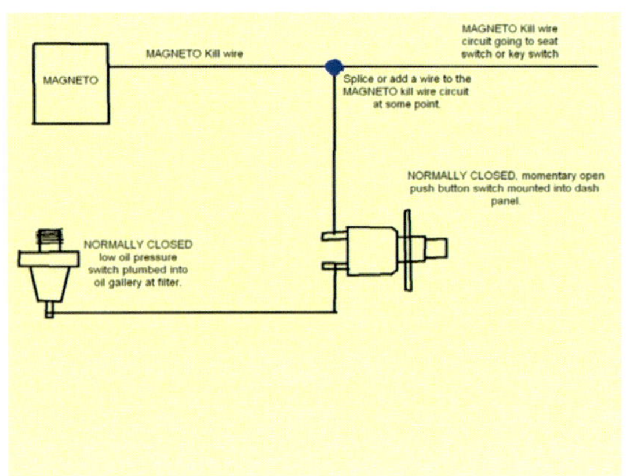

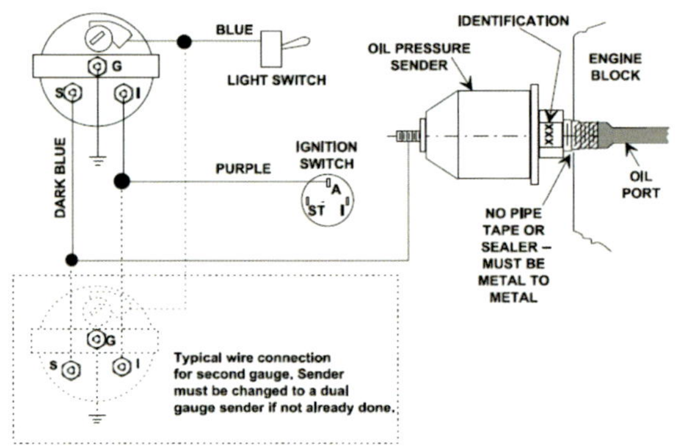

Four Stroke Engine Oil

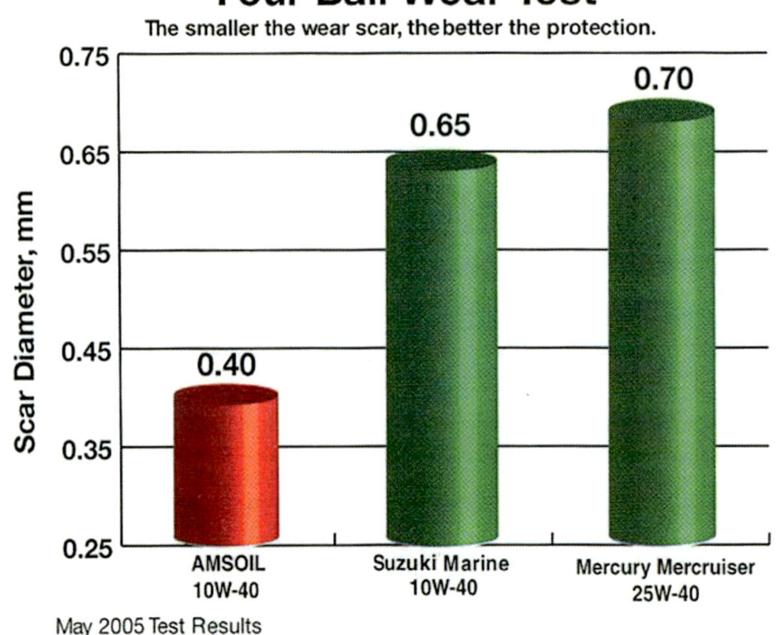

Mineral Oil

- The objective of the refining process is to isolate the desired base oils, also known as mineral oils

- The problem is that after refining operations are performed, a wide variety of chemical components remain that can affect the size and structural arrangement of molecules

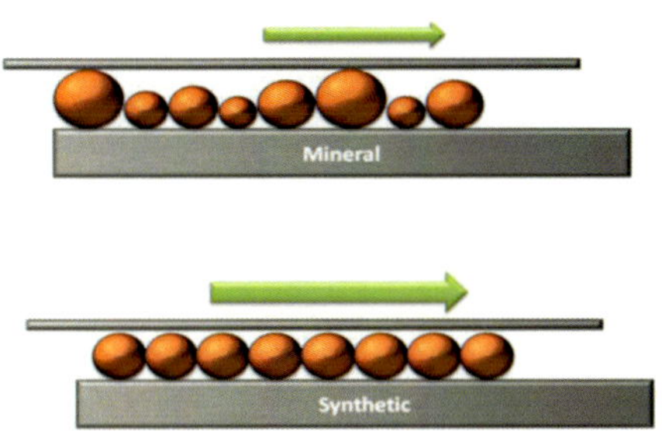

Boiling Temperatures of Mineral Oil Brake Fluids

Mineral Oil	Boiling Point	Difference
Magura Royal Blood	120 °C (248 °F)	--
Valvoline LHM+	249 °C (480 °F)	107%
Shimano	280 °C (536 °F)	12%
Juice Lubes	290 °C (554 °F)	3.5%

Mineral or Synthetic ?

Most two-stroke oils are mineral-based, but you're seeing more and more synthetics these days. Mercury's Opti-Max oil is partially synthetic

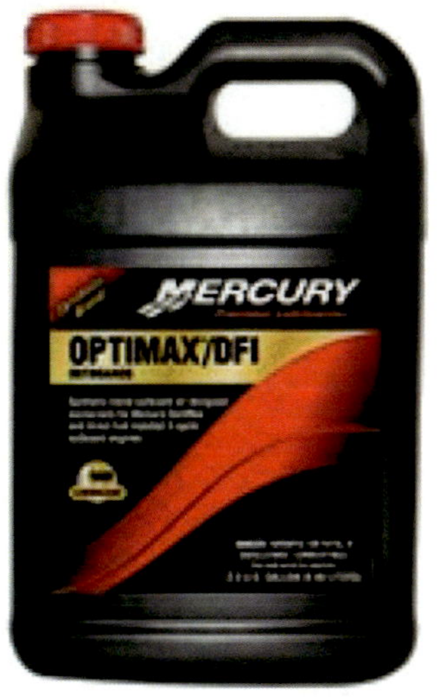

Mineral or Synthetic ?

Most two-stroke oils are mineral-based, but you're seeing more and more synthetics these days. Mercury's Opti-Max oil is partially synthetic

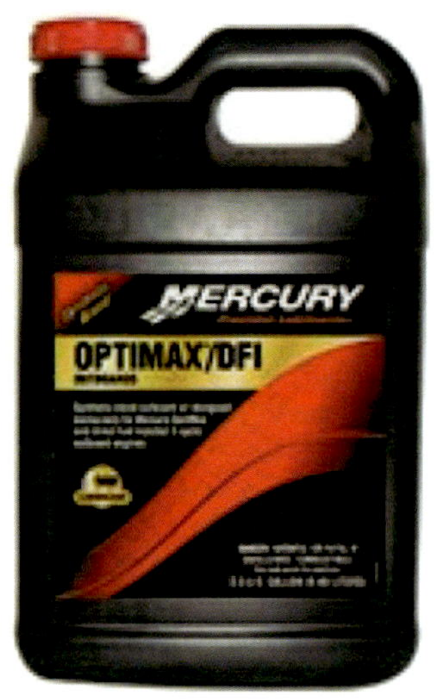

Synthetic oil

- Synthetic oils for marine engines can cost twice as much as conventional oils. But does your outboard or inboard really benefit from lab-spawned lubricants instead of those distilled

- "Synthetics flow more smoothly on cold engine start-ups than conventional oils," says Greenwood in pointing out one of the reasons for the change. "That really helps minimize engine wear."

Full Synthetics

- Full synthetics really come into play when an engine is turning 8,000 rpm and up, full synthetics are less likely to break down at high operating speeds and temperatures

- Most marine engines don't exceed 6,400 rpm, so using an expensive full synthetic amounts to a waste of money

Full Synthetic Marine Oil

- Full-synthetic oils help because they have fewer impurities that can oxidize and turn acidic when subjected to the higher temperatures inside a Gas / diesel engine with a turbocharger and circulating exhaust gases

- Mercury recommended conventional engine oil during its Cummins MerCruiser Diesel partnership (which was recently dissolved), but now recommends full-synthetic oil in its Volkswagen-based Mercury Diesel TDI power plants

What is a TC-W3

This rating is an industry certification, administered by the National Marine Manufacturers Association (NMMA), the people who operate the majority of boat shows around the country

The industry came up with a testing procedure to certify a basic and uniform level of lubrication and engine protection, and TC-W3 is the latest link in that evolution

What is a TC-W3

- Oils tested for certification by the NMMA must pass a series of grueling tests to receive the TC-W3 certification

- If an oil meets or exceeds a certain set of criteria in each of the tests, it gets a TC-W3 rating, and the containers in which the oil is subsequently sold can carry a TC-W3 emblem signifying that the oil passed the tests

Blended Oils

Many current outboard two-stroke oils are built on a mid-viscosity base oil that would be considered around 20-weight. The base oil makes up approximately 60 percent of the final product. Then a heavier base oil called brightstock is added, normally at about five percent of the total, but sometimes as much as 17 percent

Opti-Max 2-Cycle oil

"Opti-Max oil is not TC-W3-rated, so far, because we haven't ever submitted it for testing. It's far beyond the TC-W3 standard."

All of the oils that carry the <u>TC-W3</u> credentials are good; you just have to decide if some are better than others

Brightstock Additive
(2 stroke Oil additive)

- Brightstock is a great lubricity agent and helps make the oil more slippery.

- Up to 20 percent of the final blend is solvent, which acts as a carrier and helps the oil mix thoroughly with gasoline (remember that there are still a lot of pre-mix engines in circulation).

- About one percent or less is dye, so you can tell if the pre-mix gas has oil in it .The remainder is made up of additives, which may or may not contain polyisobutylene, commonly known as PIB. PIB is a synthetic oil used for both lubrication and smoke control.

Oil Grades

- The Society of Automotive Engineers (SAE) has established a numerical code system for grading motor oils according to their viscosity characteristics

- SAE viscosity gradings include the following, from low to high viscosity: 0, 5, 10, 15, 20, 25, 30, 40, 50 or 60.

- The numbers 0, 5…25 are suffixed with the letter W, designating their "winter" (not "weight") or cold-start viscosity

Total Base Number

- An oil's total base number **(TBN)** is a measure of its reserve alkalinity
- Diesel fuel contains some degree of sulfur that produce sulfuric acid that acts as a corrosive to engine components
- The higher the oil's TBN, the greater its neutralizing ability
- Most current Diesel engine oils have a TBN between 10 and 12

Total Acid Number

- The **TAN** is a companion to the TBN it measures the acidity of an oil
- The higher an oil's **TAN** number, the more acidic it is
- A high TAN can be caused by acid contamination from the environment
- Using the wrong lubricant or burning-high sulfur fuel may also result in high TAN values

OIL GRADES

Single-Grade

- Constant Viscosity to different temperatures
- The 11 viscosity grades are 0W, 5W, 10W, 15W, 20W, 25W, 20, 30, 40, 50, and 60

Multi-Grade

- Variable Viscosity. To low Temperature , low viscosity and high viscosity to high Temp.
- for example, **10W-30** designates a common multi-grade oil

Types of Exhaust Systems

- There are two basic types of marine exhaust systems, **wet** and **dry**

- A system is considered a **wet system** when the exhaust gases are mixed with the raw water at the exhaust manifold

Wet System

Most of the inboard gasoline engines works with a wet system. and all the outboard engines (two and four stroke gasoline engines) work with a wet system

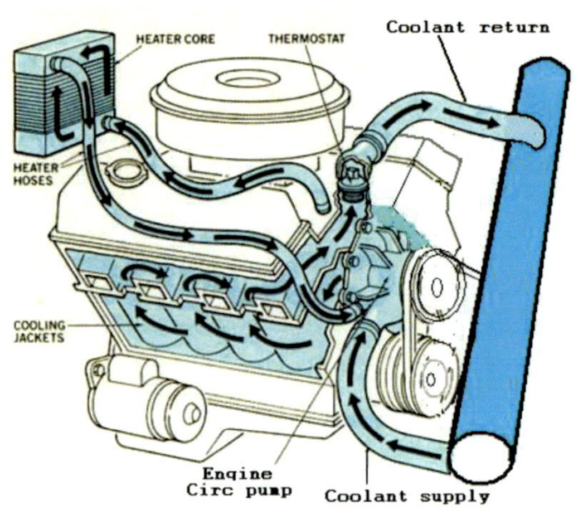

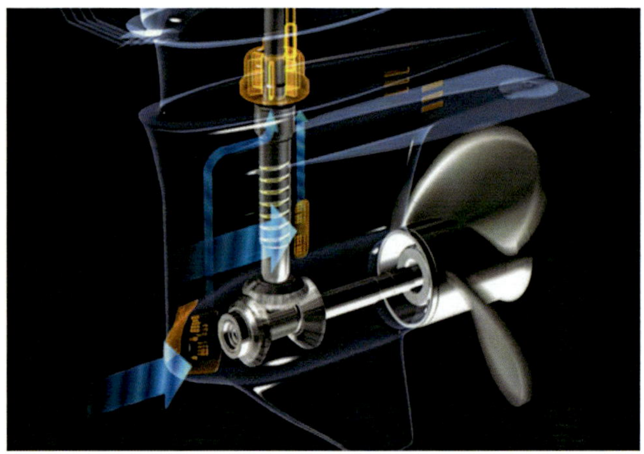

Salt Water into the Cylinders

- How the salt water can enter into the combustion chamber in an inboard engine ?

- If the engine use a closed cooling system (Coolant and Heat-exchanger), the only way is due saltwater passed into the combustion chamber through a leak in either, the raw-water cooled exhaust "manifold" or the "riser." Normally during the cranking procedure

Water Coming Back In Through the Exhaust System

Unlike automotive engine exhaust systems, marine engine exhaust systems are wet and very close to water. A marine engine's cooling water and exhaust mix at the end of the exhaust <u>elbow on the engine</u>

Water Coming Back In Through the Exhaust System

The primary causes for water to get back into the engine through the exhaust system are

1) Engine Running Conditions,

2) Boat's Waterline,

3) Water Ingestion

4) Engine Exhaust System Component. Each is described in depth following.

Engine Running Conditions

- **Ignition misfire** (fouled spark plug).
    - If an engine has a spark plug that is not firing, that cylinder will act like an air compressor and it can draw water backward into the exhaust manifold. Inspect all spark plugs, wires and the ignition system to make sure this was not the cause for water in the engine
- **Bad valves**
    - Valves that are not seating properly can draw water backward into the engine. Take engine compression. If low, grind valves and seats

Engine Running Conditions

- Engine run-on (dieseling).
    - An engine that continues to run after shutting the ignition key off can draw water backward into the exhaust manifold. Engine can run in reverse direction when it 'diesels'. If an engine 'diesels' when the key is turned off, turn key back on and increase rpm (in neutral) to 1300 for 40 seconds then return engine to idle rpm
- Sticking exhaust valve.
    - An exhaust valve that sticks in its valve guide can draw water backward. Cylinder will act like an air compressor. Repair cylinder head, as necessary

Boat's Waterline

- The normal symptom for this problem is the engine ran fine until the boater turned the engine off for a period of time. When they went to restart the engine, it was 'locked up' or it did not run smoothly
  - Swim platforms and transom plate 'pockets'
  - Exhaust elbow height too low
  - Too much weight at stern of the boat or boat was 'beached'
  - Exhaust hose/collect/muffler downward angle (engines with thru-hull exhaust only).

Water Ingestion

- Engine cooling water mixes with the exhaust at the end of the exhaust elbow to go overboard. Under certain conditions, a fine mist or droplets of water can be drawn backward into the exhaust passage of the exhaust elbow while the engine is idling

- When the engine is shut off, these droplets flow downward and collect in the exhaust manifold runners that go to the cylinder head. In saltwater areas, the water evaporates and leaves a salt crystal deposit in the runner

Engine Over the Waterline

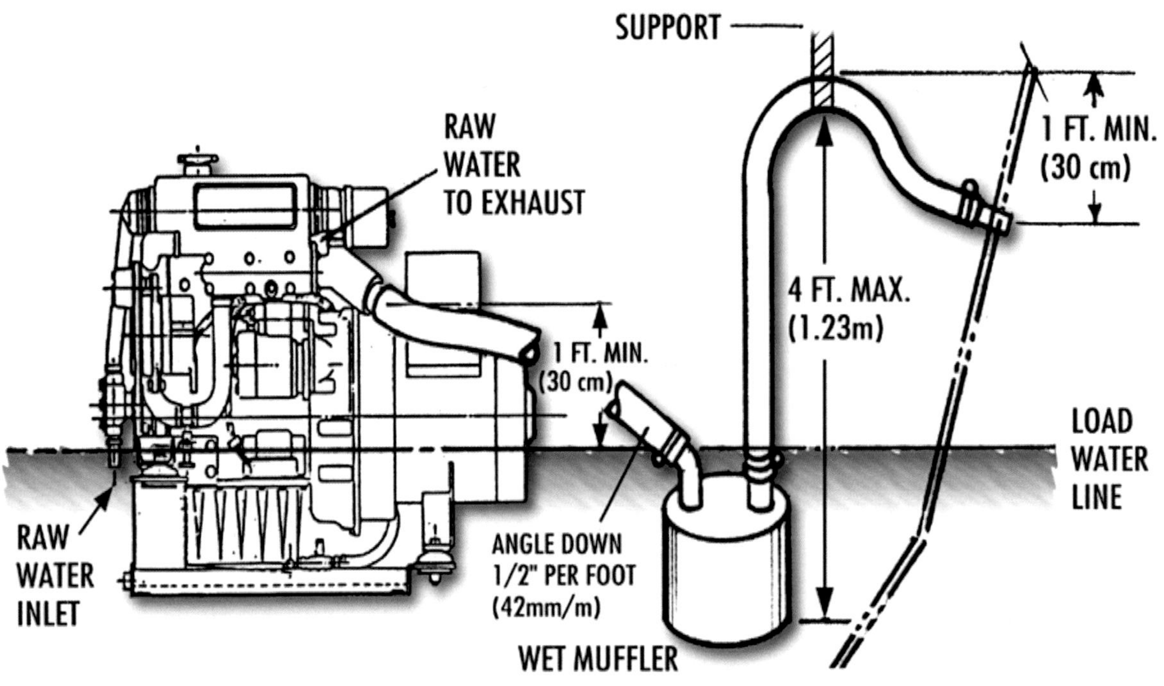

Engine Under the Waterline

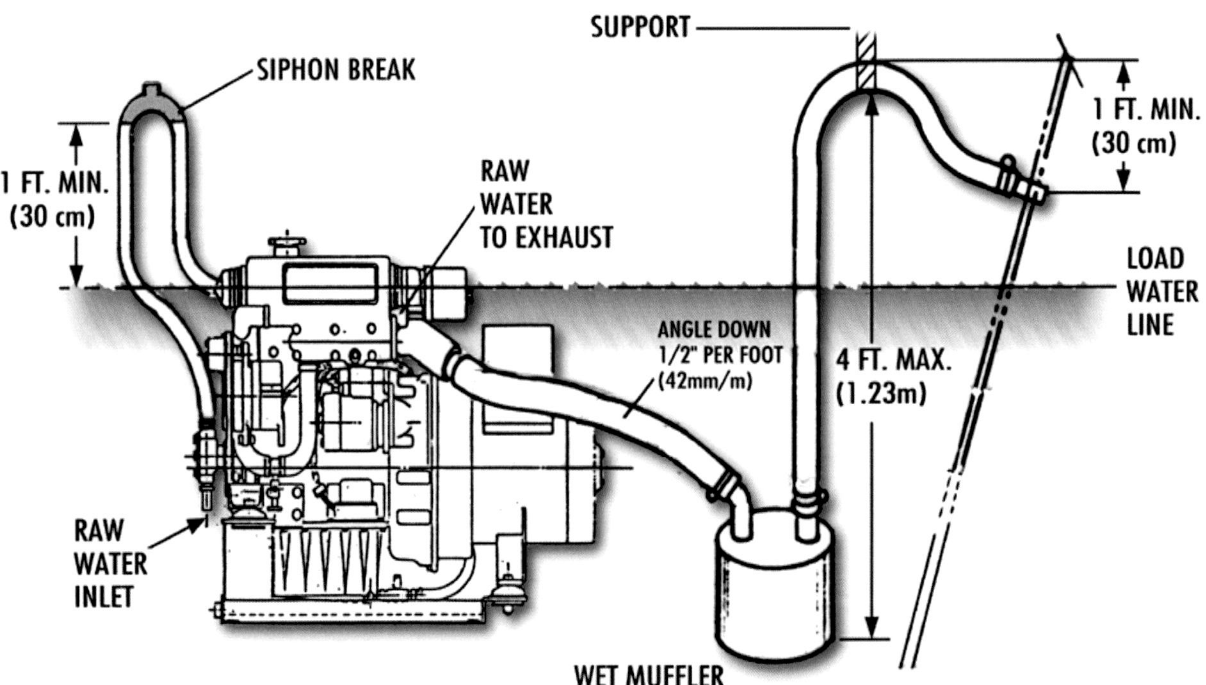

Water Ingestion

- Over time, these salt crystals will cause rust to form on the exposed surface of an open exhaust valve. When this valve sticks, it will cause more water from the exhaust elbow discharge to be drawn backward into the engine

- This condition is more likely to occur on engines that have exhaust systems that exit out the boat's hull. Sterndrive engines that have thru-prop exhaust are least likely to see this condition. Look inside the exhaust passage (at hose end of the exhaust elbow) to see if a salt or rust trail is present from that point on backward toward the manifold.

Salt Water into the Cylinders

- If the engine use a raw water cooling system (Salt water passing through the block), there are many possibilities , due that an small crack on the cylinder walls or on the head could be expanded due to the high temperatures , allowing the pass of salt water into the cylinders with a catastrophic engine failure

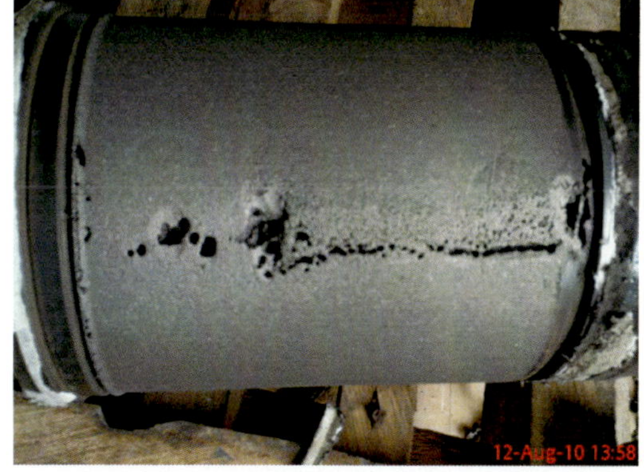

Corrosion in Raw Water System

- In an engine that is directly cooled by salt water, corrosion and rust build up inside the engine
- This restricts the water way sometimes causing the engine to operate at higher temperatures, and may corrode the internal cooling system of the engine causing engine failure

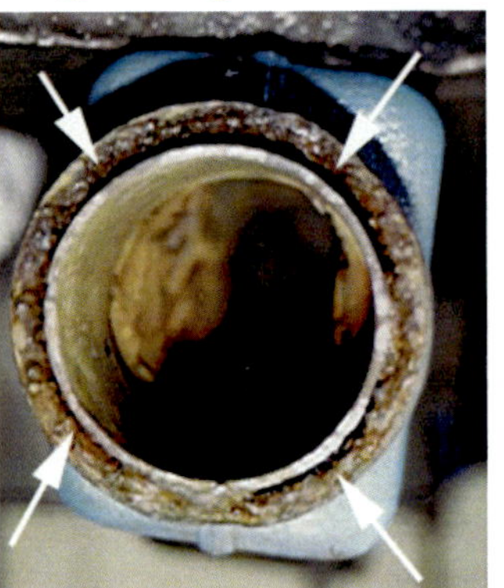

Marine Exhaust Systems

The design has to be durable and safe, resistant to corrosion and corrosive gases, and not transmit noise and vibration into the boat structure

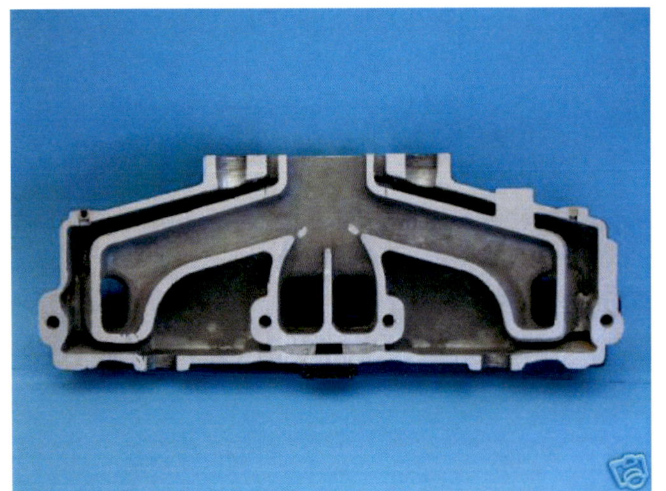

Marine Exhaust Systems

The design also needs to protect the interior of the engine from heat and to reduce back pressure that would keep the engine from running at maximum efficiency

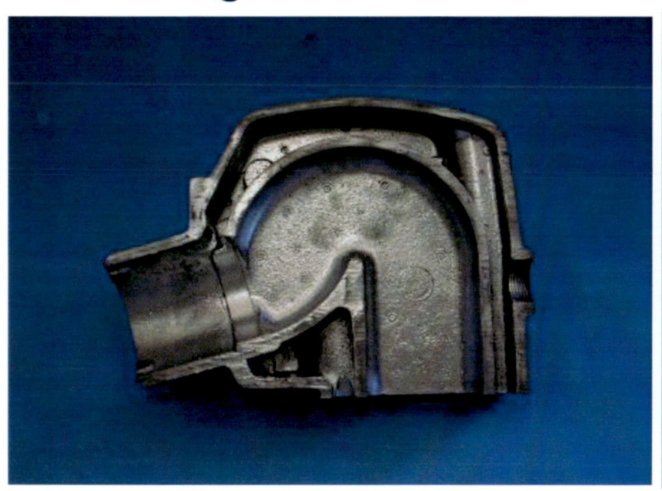

Water Cooled Manifold

Wet exhaust systems usually consist of a water cooled exhaust manifold, a riser off the manifold that lifts the exhaust up from the manifold and then down again

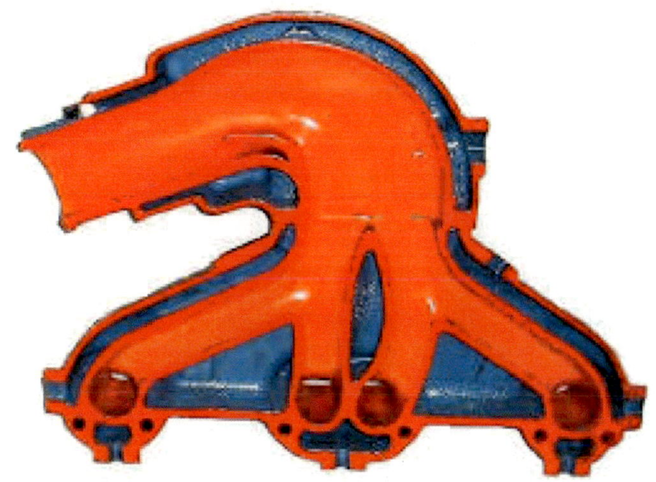

The Risers

- The riser prevents water in the exhaust system from being ingested into the engine through the exhaust manifold

- At this point the cooling water is injected into the exhaust and blown out the exhaust opening along with the hot exhaust gases. This helps to muffle and cool the exhaust

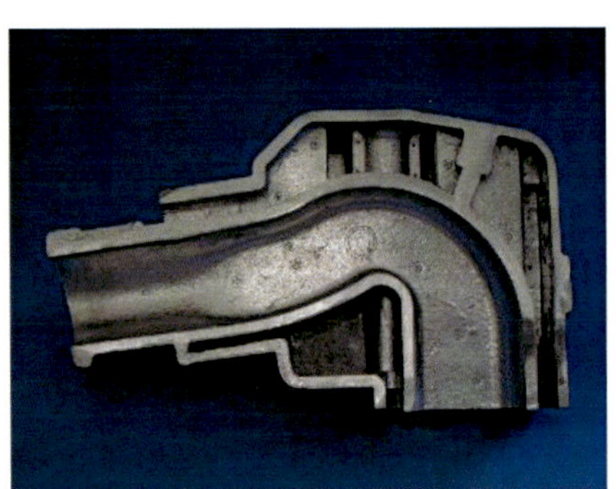

The Muffler

- After the Raiser the system may be a muffler in the system to quiet the engine even more

- In the near future there may also be a catalytic converter in the system to reduce exhaust emissions

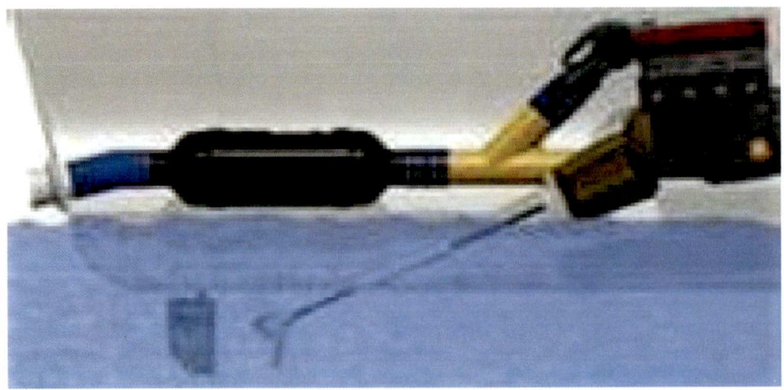

234

Water Lift Silencers

'Water lift' or 'water-trap' silencers can be positioned well below the waterline, allowing the exhaust run from the engine to have a nice, safe, steep angle

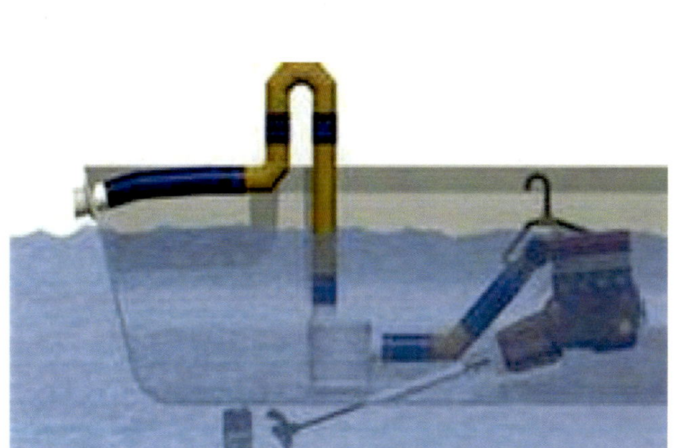

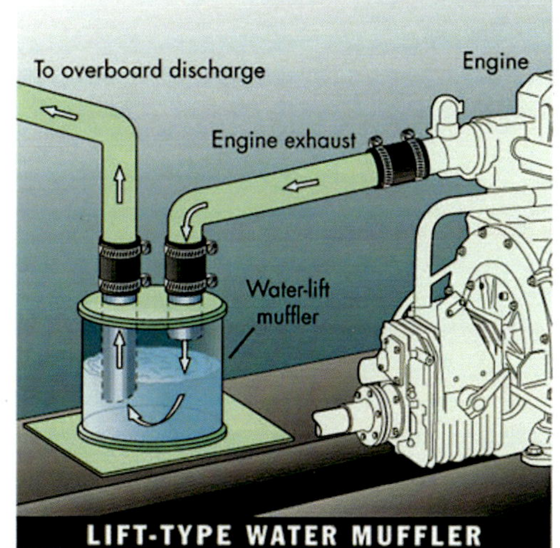

Water Lift Silencers

- You might think that such a device would rapidly fill with water and eventually block the exhaust completely, but this doesn't happen

- In a correctly designed unit, the water level stabilizes in the bottom of the chamber, just level with the bottom of the outlet pipe, with pure gas swirling around above it.

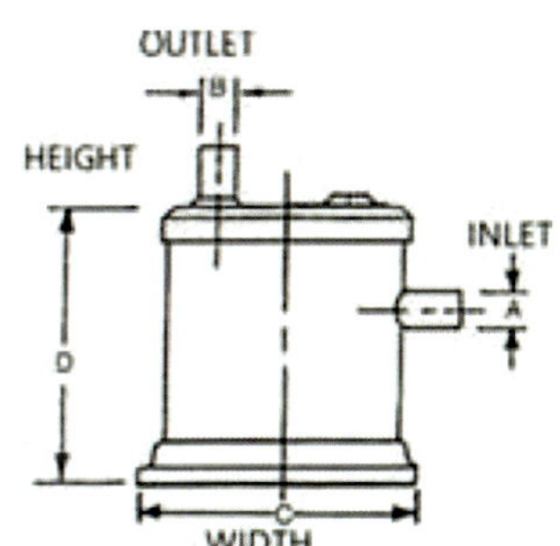

Waterlift Silencers

- Water is picked up by the gas on the way out and this further silences and cools the flow .

- Waterlift silencers should reduce exhaust noise by about 40%, compared with an unsilenced system .

- Most common is the vertical cylinder variety, which can have top inlets, horizontal or angled side inlets and a wide variety of outlets

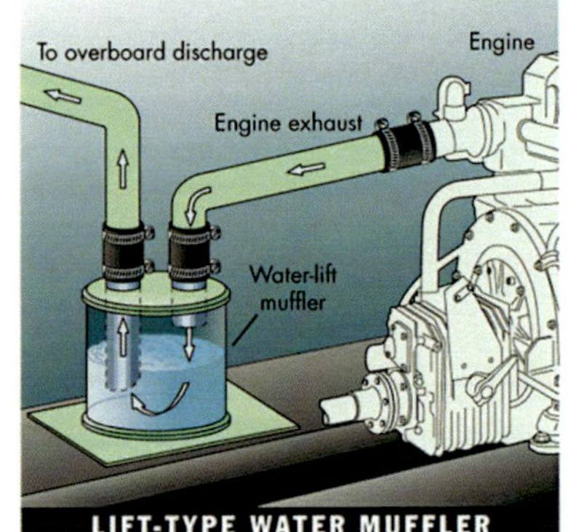

Exhaust water after the exhaust manifold

In order to avoid inconveniences with the salt water passing through the exhaust manifold , In newer designs, the salt water expelled from the heat exchanger is connected at the output of the exhaust manifold

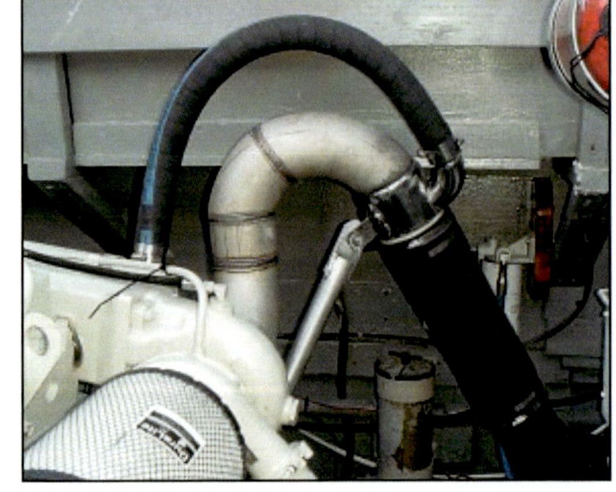

236

Dry Systems

Dry systems may or may not have water cooled manifolds. But after the exhaust exits the manifold, no water is injected into the system

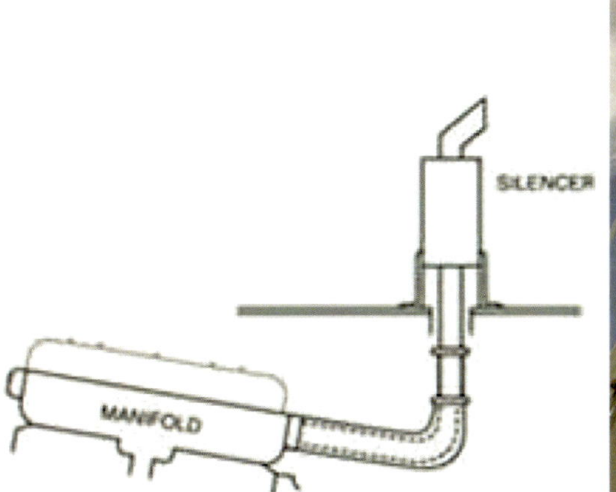

Dry Exhaust System

The exhaust pipe can then rise up vertically or go horizontally out of the boat. There is usually a muffler in the exhaust pipe

Dry Exhaust System

The pipes get very hot so usually they are wrapped with fiberglass lagging to insulate the interior of the boat from the heat and to prevent burns and fires

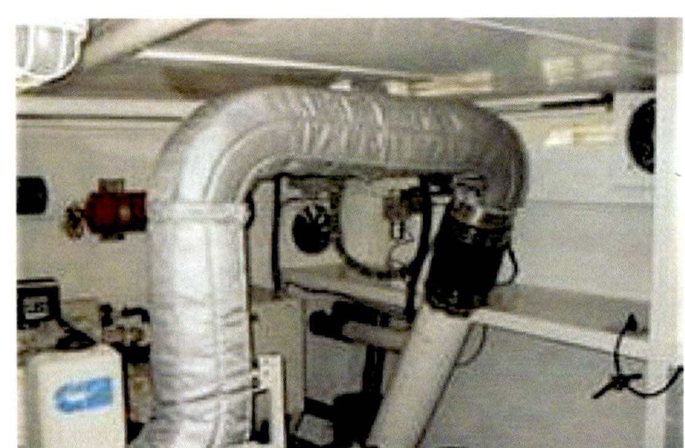 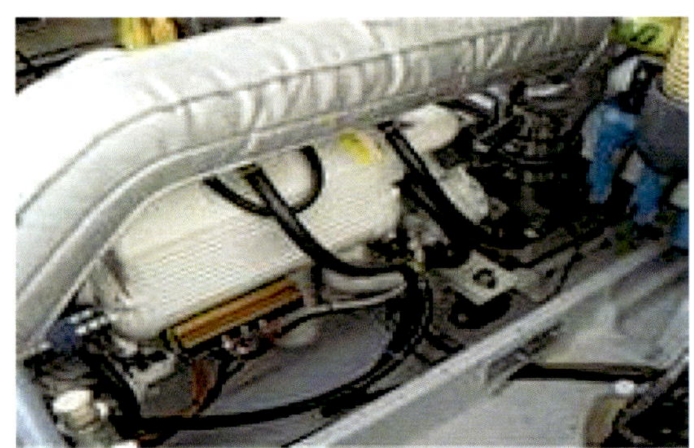

Dry exhaust Systems

Dry stacks, that is, tall vertical exhaust pipes, are commonly used on larger recreational and commercial boats to direct exhaust gases as far up and away from the boat as practical. This prevents carbon monoxide and other noxious gases from getting into the boat

Water Separator

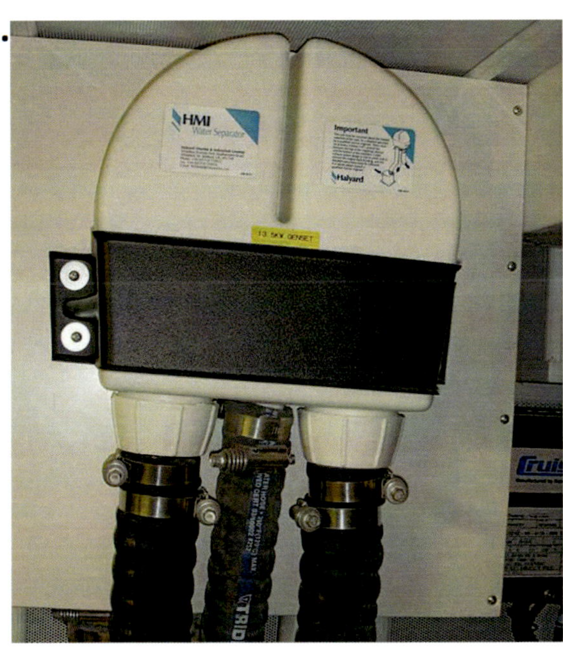

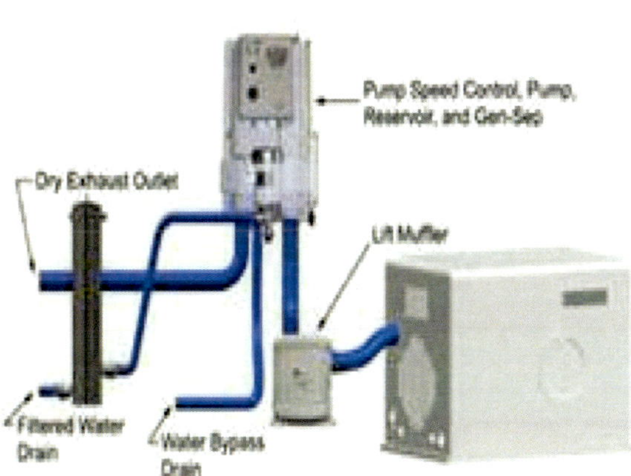

Installation with Water Separator

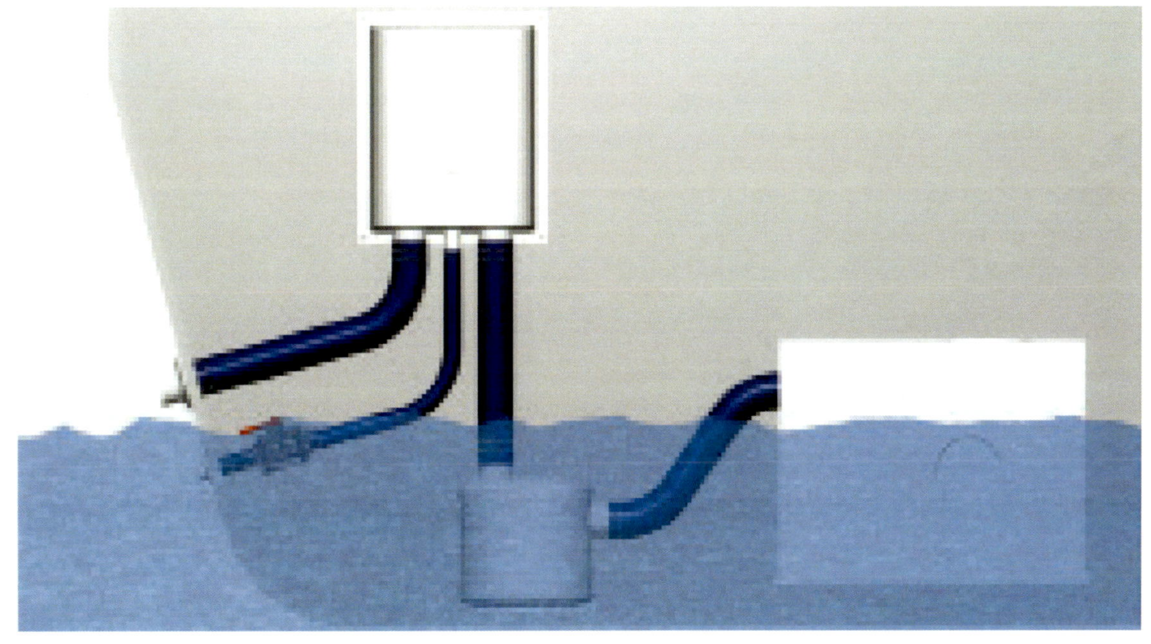

Exhaust Back pressure

- One of the most misunderstood concepts in exhaust theory is back pressure.
- Your exhaust system is designed to evacuate gases from the combustion chamber quickly and efficiently.
- Exhaust gases are not produced in a smooth stream; exhaust gases originate in pulses. A 4 cylinder motor will have 4 distinct pulses per complete engine cycle

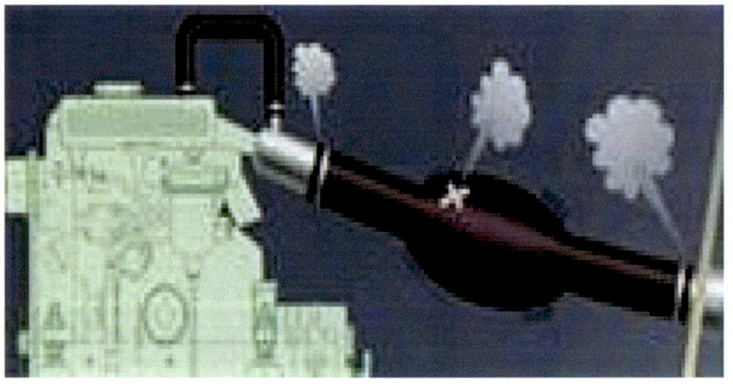

Exhaust Back pressure

- The more pulses that are produced, the more continuous the exhaust flow.
- Back pressure can be loosely defined as the resistance to positive flow - in this case, the resistance to positive flow of the exhaust stream

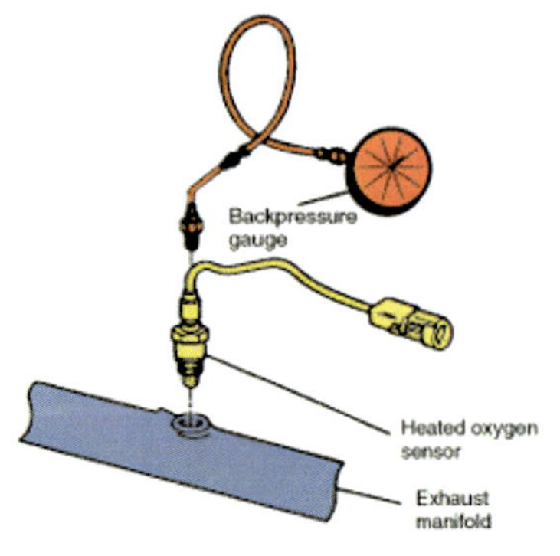

Back Pressure and Velocity

- Many people mistakenly believe that wider pipes are more effective at clearing the combustion chamber than narrower pipes.

- According with the VELOCITY concept. Here is an analogy. If you let the water just run unrestricted out of the hose it flows at a rather slow rate. However, if you take your finger and cover part of the opening, the water will spray out at a much faster rate

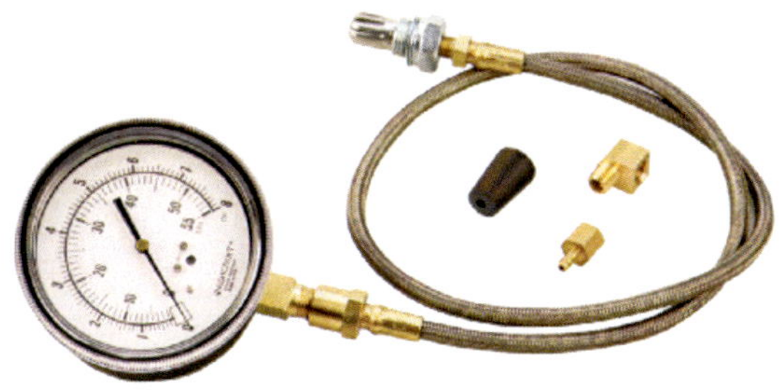

Exhaust Back Pressure

- You want the exhaust gases to exit the chamber and speed along at the highest velocity possible - you want a FAST exhaust stream

- If you have two exhaust pulses of equal volume, one in a 2" pipe and one in a 3" pipe, the pulse in the 2" pipe will be travelling considerably FASTER than the pulse in the 3" pipe

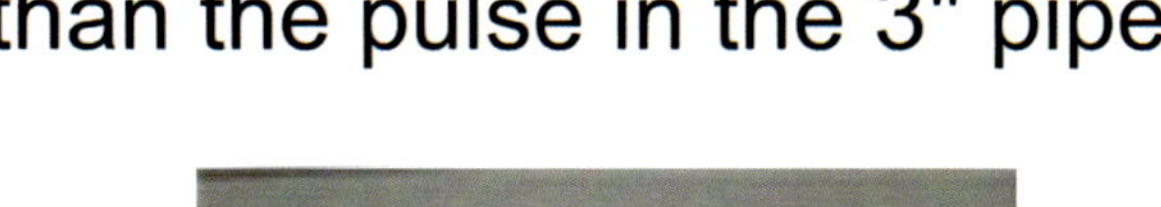
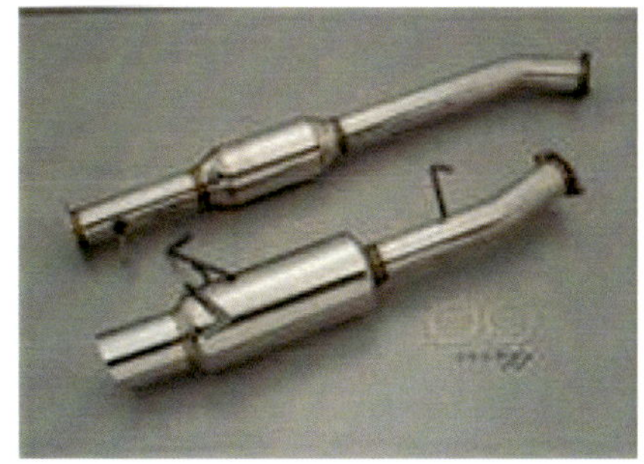

Exhaust Back Pressure

- **Back pressure** refers to pressure opposed to the desired flow of a fluid in a confined place such as a pipe
- The trick is to have a pipe that that is as narrow as possible while having as close to zero back pressure as possible at the RPM range you want your power band to be located at

Exhaust Pipe Diameter

- Exhaust pipe diameters are best suited to a particular RPM range
- A smaller pipe diameter will produce higher exhaust velocities at a lower RPM but create unacceptably high amounts of back pressure at high rpm

Back Pressure

- In four stroke engines the back pressure caused by the exhaust system (consisting of the exhaust manifold, muffler and connecting pipes) of a four stroke Diesel has a negative effect on engine efficiency resulting in a decrease of power output that must be compensated by increasing fuel consumption

- In a two stroke engine however, a certain amount of exhaust back pressure is needed to prevent unburned fuel/air mixture from passing right through the cylinders into the exhaust

Efficiency

- Dynamometer testing has proven that **"hot"** cylinders are a common cause of premature engine failure, especially in high performance marine applications

- The polishing of exhaust cavities improve scavenging of each cylinder to remove exhaust gasses that produce potentially damaging high cylinder head temperatures

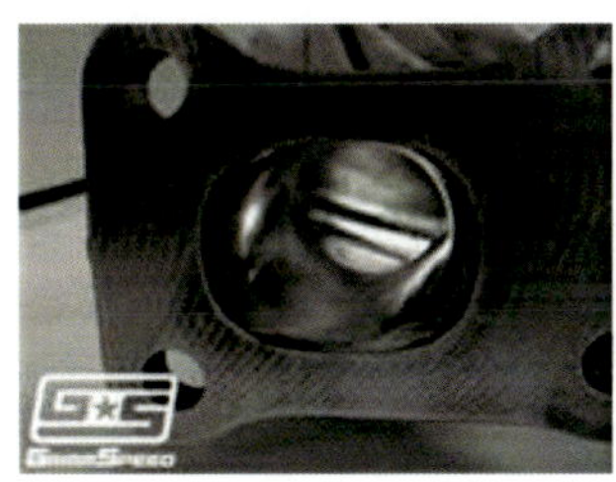

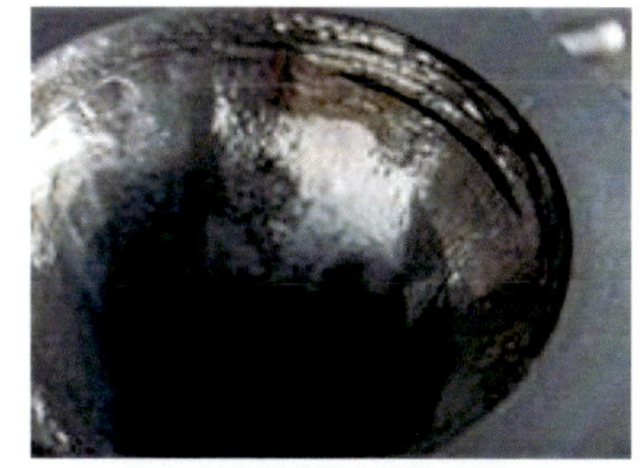

Flexible Exhaust Hose

- Hose used in wet exhaust systems shall comply with the performance requirements of UL1129

- Every exhaust hose connection shall be secured with at least two non-overlapping clamps at each end to produce a secure, liquid and vapor-tight joint

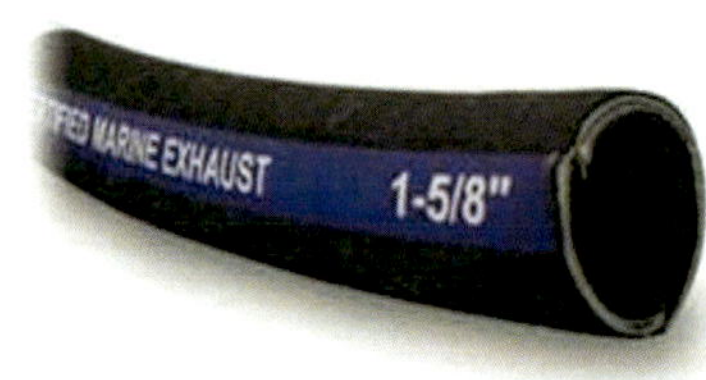

Hose and Clamps

- Clamps used for this purpose shall be entirely of stainless steel metal

- The bands shall be a minimum of one-half inch in width

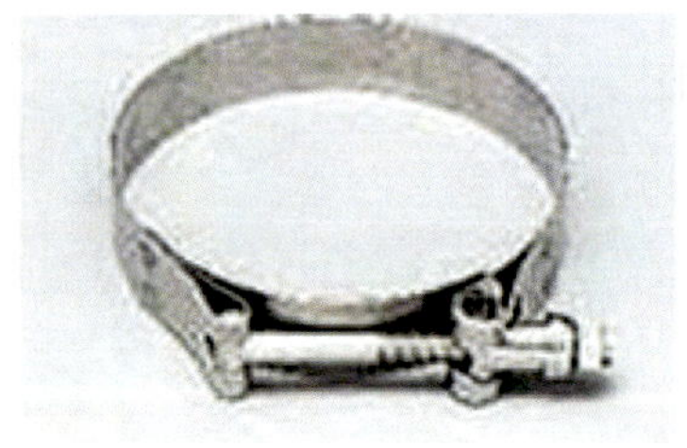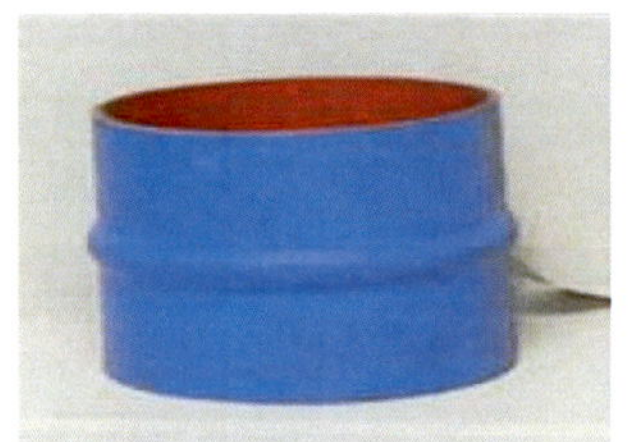

The Siphon Break

- The siphon break, or air break, is basically a small air valve which is fitted in the raw water hose line located high on the bulkhead above the engine

- This hose conveys the seawater from the engine's heat exchanger to the water injection elbow just aft of the engine exhaust manifold .

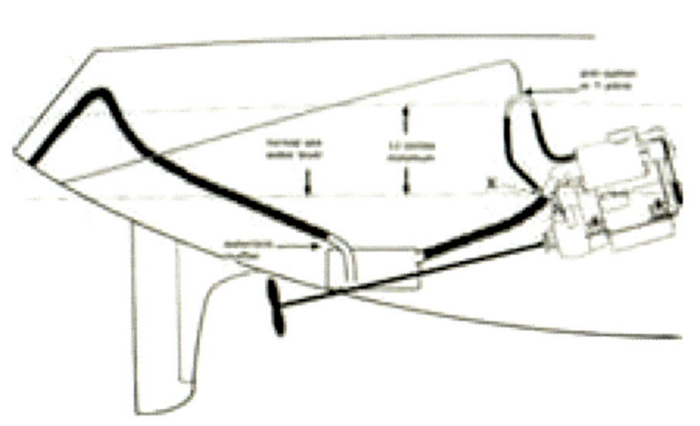

The Siphon Break

When the engine is running, sea water is drawn in via the thruhull valve [under the engine], passes through the intake strainer, the raw water pump, the heat exchanger, then up past the siphon break and down into the exhaust water-injection elbow, exiting via the exhaust muffler and hose

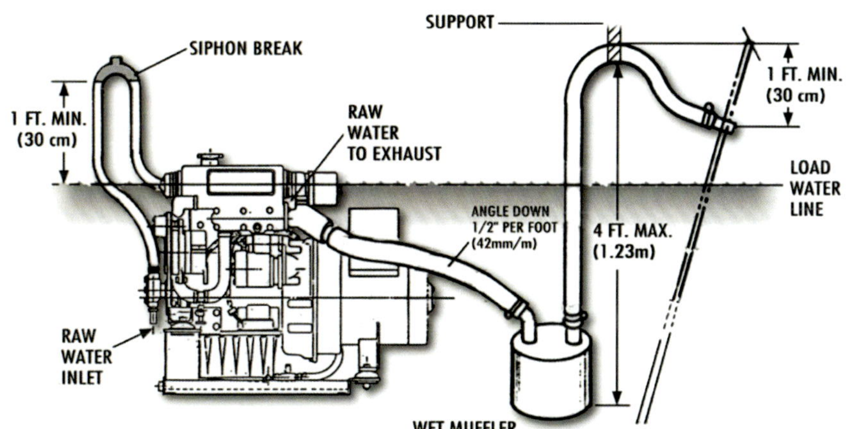

The Siphon Break

The air valve closes under water pressure with the engine running. When the engine shuts down, this valve opens and air is drawn into the hose, allowing the water within the hose to then fall away under gravity, down into the exhaust muffler

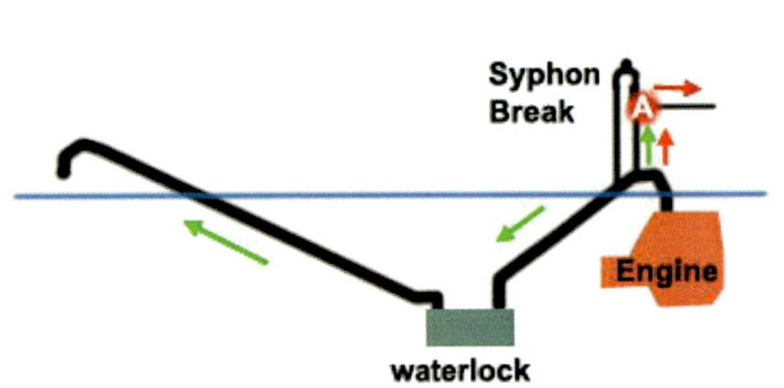

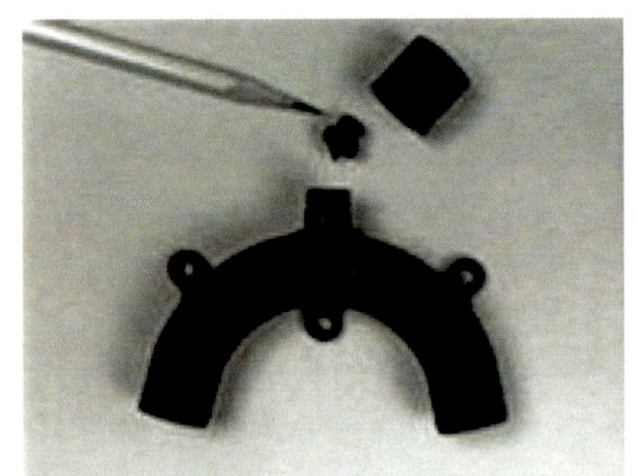

The Siphon Break

This is the critical function of the device: to prevent seawater from otherwise siphoning past the pump or being drawn in from the muffler, possibly filling the engine's exhaust manifold with sea water

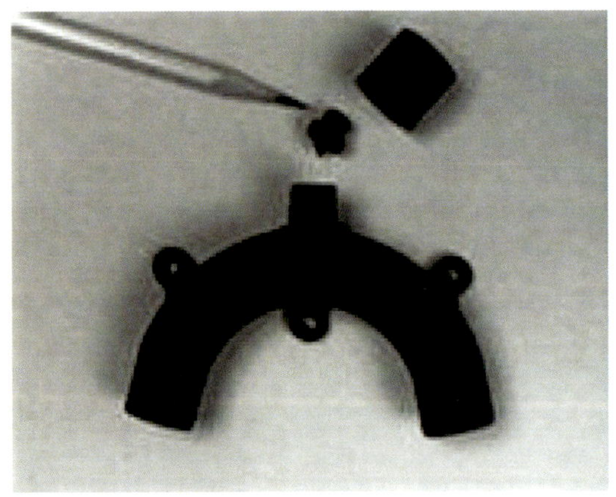

The Siphon Break

There are various types of air or siphon breaks available. The most common type consists of a lightweight ball, about the size of a pea, enclosed within a small plastic cage or basket on top [see Photo 1]. The ball presses up against a seat when water is flowing under pressure from the pump, and falls to allow air into the hose when the engine is shut down, stopping the pump

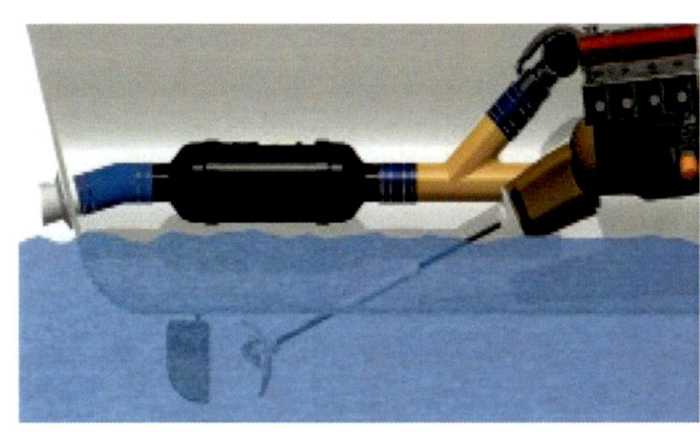

typical installation diagrams

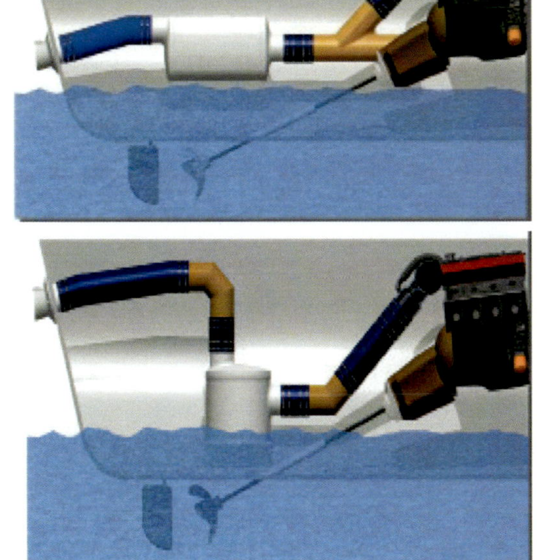

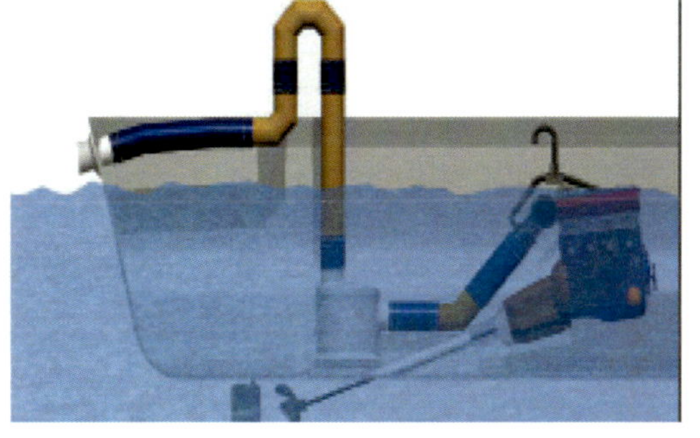